SpringerBriefs in Speech Technology

Series Editor:
Amy Neustein

More information about this series at http://www.springer.com/series/10043

Editor's Note

The authors of this series have been hand-selected. They comprise some of the most outstanding scientists – drawn from academia and private industry – whose research is marked by its novelty, applicability, and practicality in providing broad based speech solutions. The SpringerBriefs in Speech Technology series provides the latest findings in speech technology gleaned from comprehensive literature reviews and *empirical investigations* that are performed in both laboratory and *real life* settings. Some of the topics covered in this series include the presentation of real life commercial deployment of spoken dialog systems, contemporary methods of speech parameterization, developments in information security for automated speech, forensic speaker recognition, use of sophisticated speech analytics in call centers, and an exploration of new methods of soft computing for improving human-computer interaction. Those in academia, the private sector, the self-service industry, law enforcement, and government intelligence, are among the principal audience for this series, which is designed to serve as an important and essential reference guide for speech developers, system designers, speech engineers, linguists, and others. In particular, a major audience of readers will consist of researchers and technical experts in the automated call center industry where speech processing is a key component to the functioning of customer care contact centers.

Amy Neustein, Ph.D., serves as Editor-in-Chief of the International Journal of Speech Technology (Springer). She edited the recently published book "Advances in Speech Recognition: Mobile Environments, Call Centers and Clinics" (Springer 2010), and serves as quest columnist on speech processing for Womensenews. Dr. Neustein is Founder and CEO of Linguistic Technology Systems, a NJ-based think tank for intelligent design of advanced natural language based emotion-detection software to improve human response in monitoring recorded conversations of terror suspects and helpline calls. Dr. Neustein's work appears in the peer review literature and in industry and mass media publications. Her academic books, which cover a range of political, social and legal topics, have been cited in the Chronicles of Higher Education, and have won her a Pro Humanitate Literary Award. She serves on the visiting faculty of the National Judicial College and as a plenary speaker at conferences in artificial intelligence and computing. Dr. Neustein is a member of MIR (machine intelligence research) Labs, which does advanced work in computer technology to assist underdeveloped countries in improving their ability to cope with famine, disease/illness, and political and social affliction. She is a founding member of the New York City Speech Processing Consortium, a newly formed group of NY-based companies, publishing houses, and researchers dedicated to advancing speech technology research and development.

V. Ramasubramanian • Harish Doddala

Ultra Low Bit-Rate Speech Coding

 Springer

V. Ramasubramanian
PES Institute of Technology
 – Bangalore South Campus
Bangalore, India

Harish Doddala
Oracle
Redwood Shores, CA, USA

ISSN 2191-8112 ISSN 2191-8120 (electronic)
ISBN 978-1-4939-1340-4 ISBN 978-1-4939-1341-1 (eBook)
DOI 10.1007/978-1-4939-1341-1
Springer New York Heidelberg Dordrecht London

Library of Congress Control Number: 2014943671

© The Author 2015

This work is subject to copyright. All rights are reserved by the Publisher, whether the whole or part of the material is concerned, specifically the rights of translation, reprinting, reuse of illustrations, recitation, broadcasting, reproduction on microfilms or in any other physical way, and transmission or information storage and retrieval, electronic adaptation, computer software, or by similar or dissimilar methodology now known or hereafter developed. Exempted from this legal reservation are brief excerpts in connection with reviews or scholarly analysis or material supplied specifically for the purpose of being entered and executed on a computer system, for exclusive use by the purchaser of the work. Duplication of this publication or parts thereof is permitted only under the provisions of the Copyright Law of the Publisher's location, in its current version, and permission for use must always be obtained from Springer. Permissions for use may be obtained through RightsLink at the Copyright Clearance Center. Violations are liable to prosecution under the respective Copyright Law.

The use of general descriptive names, registered names, trademarks, service marks, etc. in this publication does not imply, even in the absence of a specific statement, that such names are exempt from the relevant protective laws and regulations and therefore free for general use.

While the advice and information in this book are believed to be true and accurate at the date of publication, neither the authors nor the editors nor the publisher can accept any legal responsibility for any errors or omissions that may be made. The publisher makes no warranty, express or implied, with respect to the material contained herein.

Printed on acid-free paper

Springer is part of Springer Science+Business Media (www.springer.com)

Contents

1 Introduction 1
   1.1 Lower Limit of Bit-Rate 3
       1.1.1 Using Text Accompanying Speech 4
   1.2 Vocoder Framework 6
   1.3 Clustered Codebook to Continuous Codebook 9
       1.3.1 Clustered Segment Codebook 9
       1.3.2 Vector Quantization Performance Trends and Limits 10
       1.3.3 Random Segment Codebooks 10
       1.3.4 Vector to Segment Quantization Performance
           Retention 11
       1.3.5 A Converging Viewpoint 12
           1.3.5.1 Reasoning I 12
           1.3.5.2 Reasoning II 15
   1.4 Speech-to-Speech Synthesis by Unit-Selection 18
   1.5 Alternate Perspectives for Ultra Low Bit-Rate Speech Coding . . . 21
   1.6 Applications of Ultra Low Bit-Rate Speech Coding 23
   1.7 Organization of the Book 25

2 Ultra Low Bit-Rate Coders 27
   2.1 Vector and Matrix Quantization 29
   2.2 Segment Vocoders 32
       2.2.1 Automatic Segmentation 34
           2.2.1.1 Spectral Transition Measure 35
           2.2.1.2 Maximum-Likelihood Segmentation 36
           2.2.1.3 ML Segmentation: Duration Constrained
               (ML(DC)) 38
           2.2.1.4 ML Segmentation: A Generalized Basis 40
           2.2.1.5 Syllable-like units and other segmentations 40
           2.2.1.6 Temporal Decomposition 41
       2.2.2 Segment Quantization 42
       2.2.3 Joint Segmentation Quantization 44
           2.2.3.1 Basic framework 44

		2.2.3.2	Shiraki and Honda Variable-Length Segment Quantization	45
		2.2.3.3	2-Level DP Framework for Joint Segmentation and Quantization	46
		2.2.3.4	One-Pass DP Algorithm	48
		2.2.3.5	Phoneme Recognition and Phonetic Vocoders . . .	48
	2.2.4	Segment Codebook		50
		2.2.4.1	Template Segment Codebooks	51
		2.2.4.2	HMM Segment Codebook	53
	2.2.5	Duration Modification		53
	2.2.6	Residual Parameterization and Quantization		55
	2.2.7	Synthesis		56
2.3	R/D Optimal Linear Prediction			57
	2.3.1	Prandoni and Vetterli R/D Optimal Linear Prediction		57
	2.3.2	Variable-to-Variable Length Vector Quantization		60
	2.3.3	Multigrams Quantization		61
	2.3.4	Distortion Constrained Segmentation		61
2.4	HMM Based Recognition-Synthesis Paradigm			62
	2.4.1	HTS Based Framework		63
	2.4.2	Speaker Adaptive HMM Recognition-Synthesis		64
	2.4.3	Ergodic HMM Framework		65
	2.4.4	Ismail and Ponting HMM Based Vocoders		66
	2.4.5	Formant Trajectory Model Based Recognition-Synthesis		67
2.5	ALISP Units and Refinements			67
	2.5.1	Basic ALISP Framework		68
	2.5.2	Re-segmented Long Synthesis Units		69
	2.5.3	Short Synthesis Units by Dynamic Unit Selection		70
	2.5.4	Pre-selection of Units		70
	2.5.5	Noise Robustness		71
2.6	Speaker Adaptation in Phonetic Vocoders			71
2.7	Unit-Selection Paradigms			72
2.8	Performance Measures for Segment Quantization			75

3 Unit Selection Framework 79

3.1	Lee-Cox Single-Frame Unit Selection Quantization	81
	3.1.1 An Alternate '5 ms Segment' Single-Frame Unit-Selection Algorithm	82
3.2	Lee-Cox Segmental Unit Selection Quantization	83
3.3	Run-Length Coding and Effective Bit-Rate	88
3.4	Sub-optimality of Lee-Cox Segmental Unit-Selection Algorithm	90

4 Unified and Optimal Unit-Selection Framework ... 91
- 4.1 Unified Unit-Selection Framework ... 92
  - 4.1.1 Proposed One-Pass DP Algorithm ... 94
    - 4.1.1.1 Comparison with Lee and Cox Single-Frame and Segmental Unit-Selection ... 96
  - 4.1.2 Experiments and Results ... 98
- 4.2 Comparison with Lee and Cox Suboptimal Segmental Unit Selection ... 100
- 4.3 Comparison with VQ, MQ, VLSQ ... 103
  - 4.3.1 Experiments and Results ... 104
- 4.4 Conclusions ... 107

5 Optimality and Complexity Considerations ... 109
- 5.1 Complexity of 1-Pass DP Optimal Unit-Selection Algorithm ... 110
  - 5.1.1 Local Distance Calculations Cost ... 111
  - 5.1.2 Within-Unit Recursion Cost ... 111
  - 5.1.3 Cross-Unit Recursion Cost ... 111
- 5.2 Complexity of Lee-Cox Segmental Unit-Selection Algorithm ... 112
  - 5.2.1 Distance Calculations ... 113
  - 5.2.2 Recursion Calculations ... 114
- 5.3 Comparison of 1-Pass DP and Lee-Cox Segmental Unit-Selection ... 114
- 5.4 Optimality-Complexity Tradeoff ... 115
- 5.5 Proposed n-Best Lattice Search ... 116
- 5.6 Conclusions ... 123

6 No Residual Transmission: Joint Spectral-Residual Quantization ... 125
- 6.1 Joint Spectral-Residual Quantization in Lee-Cox Unit-Selection Framework ... 127
  - 6.1.1 Joint Spectral-Residual Quantization ... 129
  - 6.1.2 Experiments and Results ... 133
- 6.2 Joint Spectral-Residual Quantization in an Optimal Unit-Selection Framework ... 136
  - 6.2.1 Joint Spectral-Residual Quantization in 1-Pass DP Framework ... 137
  - 6.2.2 Experiments and Results ... 140
- 6.3 Conclusions ... 143

References ... 145

Chapter 1
Introduction

Among the various speech and language technologies (such as automatic speech recognition, speech synthesis, speaker recognition, language and accent recognition, etc.), it can be easily said that 'speech coding' is the most successful, both in terms of how well the underlying sciences of information theory, quantization, source coding, channel coding, etc. apply to the real world signals, as well as in terms of how well such an application of the different theories to realize actual speech coding systems has led to their seamless and indispensable permeation into a wide range of digital communication scenarios. This is easily borne by the fact that speech coding is also the only speech technology that has reached a level of practically deployable applicability, making it possible (and hence, requiring) to be standardized, as reflected by a gamut of standards (such as prescribed by various standards organizations such as ITU-T, ETSI, MPEG, INMARSAT, DoD, NATO, TIA, etc.) in communications technologies spanning a wide range of operating conditions, each marked by specific bit-rates and associated quality of the coded speech, along with other speech coding operational parameters such as delay, complexity, robustness, etc. [C95].

While alluding to the success of a wide range of speech coders operating at various bit-rates and speech quality, governed by standards and deployable levels of maturity, we note that there is one interesting range of bit-rates, namely 1 Kbps and less, down to possibly 100 bps, that has not received as much attention or success as the other ranges, reflected in the fact that this range is yet to be standardized with practically deployable coders, but which is nevertheless marked by a host of highly specialized quantization techniques, all striving towards realizing acceptable speech quality in this challenging lower end of bit-rates.

This book will be concerned with this range of bit-rates, referred to as 'ultra-low' bit-rate speech coding, with this range posing a challenging problem as it deals with an extreme compression of speech from what is defined as the reference bit-rate of 128 Kbps for un-coded speech. This reference bit-rate corresponds to the output of an analog-to-digital converter with a linear PCM of telephone speech (band limited to 300–3,400 Hz) sampled at 8 kHz with a resolution of 16 bits/sample and is considered to yield the highest possible quality for digitally represented speech termed 'broadcast' or 'commentary' quality [F79].

In general, speech coding is primarily concerned with encoding this reference speech signal with as low a bit rate as possible while maintaining acceptable speech quality. Such bit-rate reductions (starting from the reference 128 Kbps) are naturally obtained at progressive reductions in the quality and the various coding techniques till date are categorized depending on the operating bit-rates and the quality of speech that the techniques can offer. The main categories, in terms of the quality of speech, are toll quality, communications quality and synthetic quality. The corresponding bit-rates are 16–64 Kbps, 2.4–16 Kbps and 2.4 Kbps and below [F79, Sp94, KP95b, K05]. Yet another categorization of these bit-rate ranges are: High bit rate coders (>16 Kbps), Medium bit rate coders (4–16 Kbps), Low bit rate coders (1–4 Kbps) and Very low bit rate coders (<1 Kbps). This book is concerned with the last category, namely, <1 Kbps, also referred to as "Ultra-low bit-rate coding", and more particularly at the lower ends of this range, e.g. down to 100 bps. Due to the extremely low bit-rates involved, the objective of these coders is to ensure "intelligible" speech quality while also potentially preserving the speaker-identity.

In this chapter, we limit our discussion to providing the following perspectives that allow us to view such ultra low bit-rate ranges.

(a) How ultra low bit-rates are indeed possible from qualitative considerations of the lowest achievable bit-rates, also referred as 'linguistically motivated' limits.
(b) How this is made possible within a vocoder framework; specifically considering the LPC-10 vocoder framework, we see how the definition of progressively longer units of quantization leads to the generalized framework of 'segment vocoder' that leads to progressive reduction in quantization bit-rates even while retaining the LPC-10 coder quality or better.
(c) How the notion of a large continuous codebook in the place of a clustered variable segment codebook in a segment vocoder framework leads to a break away from conventional segment quantization notions and to unit-selection based quantization frameworks, allowing the coder to potentially reach speech quality close to the 1 dB spectral distortions corresponding to transparent quality speech.
(d) How the unit-selection based segment quantization framework can be seen to be a speech-to-speech synthesis baseline of a text-to-speech synthesis framework, thereby promising quality equivalent to that of current standard high-quality single speaker TTS or better, even while having the potential to become speaker-independent by adopting means of speaker adaptation and voice conversion principles currently in vogue in text-to-speech synthesis systems.

In the following, we expand on each of these perspectives, which together characterize the basic principles driving a generic system operating in the ultra low bit-rate ranges, and particularly in the class of systems based on unit-selection based segment quantization that constitute the central theme of this book.

1.1 Lower Limit of Bit-Rate

A possible lower limit of bit-rate, also called as a 'linguistically motivated' rate [KP95a], can be derived based on the criteria that it is sufficient to transmit the phonetic content of a speech signal, which is then used to reconstruct the speech signal at the receiver by synthesizing the sounds of each phone while also concatenating them appropriately. This is shown schematically in Fig. 1.1. Such a scheme involves transcribing a given input speech signal into a sequence of the constituent phonemes of the language of the speech signal, such as can be done using a phoneme recognition system and transmitting the phoneme indices. At the receiver or the decoder, the phonemes can be converted to speech using any of the several well established speech synthesis techniques, such as rule-based synthesis, formant vocoder, or even waveform concatenation, such as the PSOLA technique. This is essentially a very coarse signal-to-symbol encoding system at the transmitter, with each symbol being one of the several phonemes in a language, and transmission of the index of the symbol in binary form, which is then used at the receiver to concatenate and synthesize the phonemes in a sequence to realize the synthetic speech.

The total bit-rate for such a system is simply IR bps, where $I = -\sum_{i=1}^{N} P_i \log P_i$ is the average bit allocation required for the phone set of N phones, and R is the number of phonemes per second under a normal speaking rate. For English language, it is reasonable to use $N=42$. Given the relative probabilities P_i of the phonemes of English (see for instance, the distribution obtained in [PD89]), we get an estimate of $I=5$ bits per phoneme and with $R=10$ phonemes/s, the effective bit-rate is 50 bps. This can be seen to be a reasonable limit even if one considers a uniform coding scheme of each of the 40–60 phonemes in any language, using an indexing scheme of $\log_2 64$ bits/phoneme, (with 64 serving as the upper bound of the number of phonemes in a language), resulting in a 60 bps lower limit of the bit-rate.

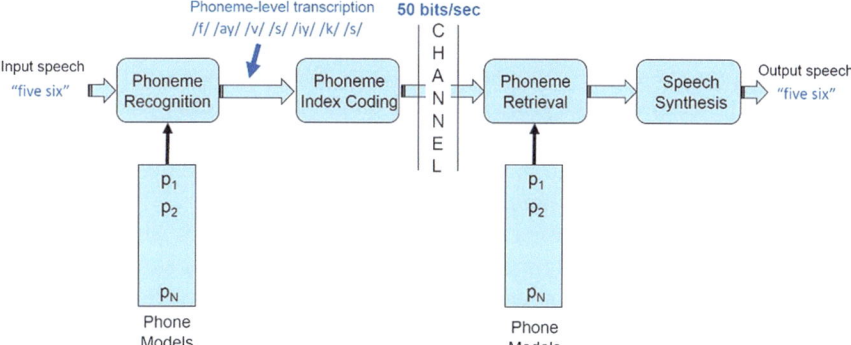

Fig. 1.1 Schematic of an ultra low bit-rate speech coder at its minimal operating limit

Thus, a bit-rate of 50–60 bps can be considered as a lower limit for the bit-rate required for a speech coder considering that the main content of speech can be given as a sequence of phonemes. Note that such a scheme does not consider quantization or transmission of other important parameters, necessary for naturalness of the coded speech, such as prosody (mainly in terms of phone duration, intonation and intensity) and speaker identity. Therefore, the speech synthesized at the receiver sans all these additional information that are crucial for natural speech, but nevertheless adequate in conveying the primary phonetic content of the speech, thereby ensuring that information akin to textual content is communicated. Such a bare-minimum skeletal system does indeed serve as a limit which ultra-low bit-rate speech coding can strive to reach while ensuring intelligibility.

Note that additional bits are required over and above the 50–60 bps estimated as above, to represent (a) prosodic information that is much needed to render the speech natural, devoid of which the synthesized speech with only the phonetic content sounds monotonic and unnatural, as well as (b) any means of representing speaker information to be able to retain the input speaker identity, even if it is known how this can be extracted from the input speech signal and incorporated back into the synthesized speech, other than perhaps using parameters such as vocal tract length, vocal tract configuration, higher formants values, bandwidths, etc., that would broadly carry such speaker-specific information (see for instance the brief discussion on this in Sect. 2.6).

1.1.1 Using Text Accompanying Speech

The above discussion essentially defined the limiting case of ultra low bit-rate speech coding as 50–60 bits/s arising from transmission of the phonetic units transcribed from the input speech, with the coded speech reconstructed by driving a synthesizer at the decoder using the phonetic information, with additional bits being required to code for information carrying prosody, speaker identity and emotion. We highlight here three different work [BD94, VB97, LC99], which represent a kind of copybook style of realizing the above 'lower-limit' baseline framework involving a transmission (or storage) of phonetic unit (indices) with additional prosody/speaker/emotion information getting incorporated into a TTS system at the decoder driven primarily by the phonetic unit stream. These represent an unusual work in the sense that they explored the question of how the text information corresponding to the input speech, if available in some unique applications, can be made use of in realizing such low bit-rates.

These work essentially used the text to extract phonetic or phone-like (e.g. segmental or sub-phone) information, which are transmitted (or stored) to further drive a TTS to realize the synthesized speech as a close reconstruction of the input speech, and where the input speech signal (to be stored or transmitted at low rates) is actually used to derive the prosodic, speaker and/or emotional information that get incorporated into the TTS system for prosody modification or in a post-processing in a speaker/emotion converter system. It can be noted that in these work, briefly outlined below, the problem was not posed so much as coding the input speech, as much as viewing the input speech signal as providing the additional prosody/speaker/emotion information that can enrich the TTS output (driven by the accompanying text) towards realizing TTS synthesized speech that is more natural and a close copy of the input speech.

These early attempts [BD84, VB97, LC99] used accompanying text transcript assumed to be available in certain applications (such as talking book, tele-text, multi-media, etc.). Among these, the earliest work of [BD84] treats the problem of realizing low bit-rate speech coding as one where the text provides the segmental information (in the form of target sub-phones and transition sub-phones derived from the text after phonological analysis) with the pitch of the input speech copied on to the frames of the sub-phones during LP synthesis (with the mapping between the input speech frames' pitch and the sub-phone sequence frames obtained by warping the LP parameter sequence of the sub-phone sequence derived from the text and the input speech via dynamic time warping) yielding an effective bit-rate of 420 bits/s and speech comparable to conventional LPC coded speech. This is essentially a segment vocoder set in a LPC-10 framework, but with the sub-phone units derived from text, rather than the input speech. Further, this work reduced the bit-rate to 100 bits/s (60 bits for coding the segmental aspect, i.e., sub-phone units derived from text and 40 bits for prosodic aspect), by using a dictionary of 32 pitch patterns and 32 duration patterns which are used to quantize an input pitch contour and duration. It can be easily noted that such a system becomes a fully speech based coder, i.e., without the text, if it were to only derive the sub-phone units from the input speech.

The proposal-like work of [VB97] (which did not report any actual results, but which is nevertheless an important contribution in setting the basis of the proposed framework), approached the problem of low bit-rate speech coding as one which can exploit the accompanying text (targeting the same specific application as in [BD94], namely the talking book – requiring compression of the speech to allow efficient storage) to drive a TTS at the 'expander', but which also uses prosodic, speaker and emotional information from the speech signal, either to modify the prosody of the TTS or as input to an explicit, post-processing module to perform speaker and emotion convertor on the speech synthesized by the TTS using only the text information. The encoder processes speech and text together, deriving speaker and emotion information by comparing the speech synthesized by TTS (at the encoder) with the actual speech, and dynamic 'conversion' control information is determined as an enrichment of the text stream, where the conversion control data carries prosodic and speaker specific information, and is stored along with text for storage compression purposes, together constituting highly reduced storage in comparison to the original speech; this is further used in an 'expander' to use the text to drive the TTS followed by a conversion of speaker characteristics and emotion using the stored conversion control data stream. Note that this method realizes a compression of the original speech which is not stored, but which is used only to derive the conversion data that enriches the TTS output during the expansion stage.

Yet another work, also making use of accompanying text is that of [LC99], though here the dependency on text is to derive the phonetic transcription after force-aligning the text with the corresponding speech, with the phonetic transcription being transmitted for further driving a TTS at the receiver. While this work was followed up further towards a more elaborate and rigorously formulated unit-selection framework [LC01, LC02], the emphasis of this particular precursor work of Lee and Cox in [LC99] was more in establishing the f_0 contour coding in a rate-distortion framework, wherein a piece-wise linear approximation is used to code f_0 in a way to minimize the bit-rate while maintaining the f_0 error below a specified threshold. The system realized a 300 bit/s rate and it should be noted that the phonetic transcription derived and transmitted from text can as well be derived from the input signal using a phoneme-recognition as has been actually done in various other systems [MCRK77, SKMKZ79, SKMS80, PD89, IP97, OPT00, T98, H03, MTK08] (as noted in Sect. 2.2.3.5).

1.2 Vocoder Framework

The landscape of speech coders can be effectively characterized in terms of quality vs. bit-rate performance as shown in Fig. 1.2. Specifically, we see that the entire range of speech coders can be broadly categorized into waveform approximating coders, parametric coders and hybrid coders, with each category having its best operating bit-rates and quality. A detailed treatment of these categories with their

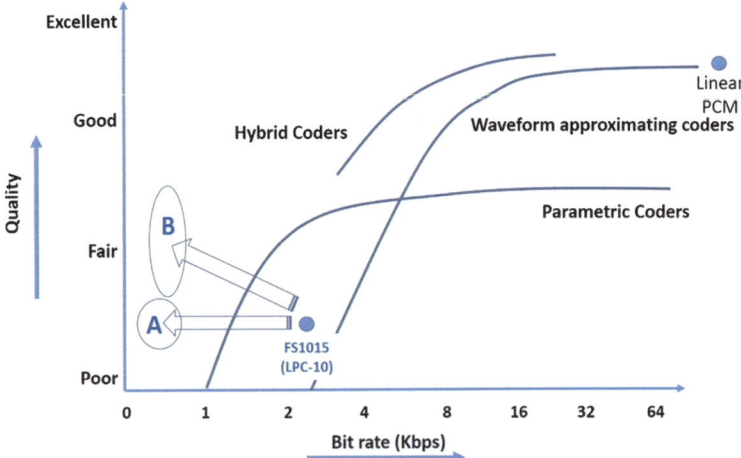

Fig. 1.2 Rate-distortion (bit-rate vs. quality) of LPC-10 vocoder and (Rate, Distortion) regions targeted by segment vocoders: Region A retaining quality of LPC-10 at ultra low rates and Region B with higher quality considering potentially low spectral distortions possible with segment quantization with very large segment codebooks, as well as better excitation modeling and synthesis methods

1.2 Vocoder Framework

individual coders is beyond the scope of this book, but can be easily looked up from several sources [F79, Sp94, KP95b, K05, M08, BSH08].

An important pivotal point in the progress towards ultra low bit-rate coding, as will be dealt with in this book, is the coder marked FS1015 on the 'parametric coders'. This is the well established linear-prediction based vocoder, also referred to as LPC-10 [T82], operating at 2.4 Kbps with a subjective speech quality considered acceptable, though sounding machine-like due to the categorical decision of voiced and unvoiced speech, with occasional perceivable annoyance due to incorrect voicing/unvoicing decision. This vocoder, nevertheless occupies a pivotal position in the sense that it led to progressive reduction in bit-rates from 2.4 Kbps all the way down to 100–300 bps by employing quantization schemes that defined increasingly longer units of quantization. Starting with scalar quantization of the linear prediction coefficients (actually, the reflection coefficients) in LPC-10, a series of quantization techniques evolved, moving from the scalar quantization to vector quantization (VQ) of the linear-prediction coefficients taken as a vector (e.g. 800 bps [W82]), further to matrix quantization (MQ) of a block of linear prediction vectors (e.g. 300 bps [WJC83, TG85]), and to the more generic segment quantization of variable length segments of linear prediction vectors (e.g. 150 bps [SH88]), finally leading to a variety of 'segment quantization' techniques that differed in the way the variable length segments were derived (automatic segmentation techniques), modeled (segments, HMM, etc.), clustered to yield variable length segment codebooks (such as random codebooks, clustered codebooks, etc.), and quantized (segment matching techniques such as dynamic time warping, HMM likelihoods with Viterbi alignment, space sampling, vector interpolation, etc.), and in the way the segments were put together at the receiver after appropriate duration normalization to match the duration of the input segments, along with other prosodic modifications including pitch and gain of the synthesized speech within a linear prediction synthesis framework. These are referred to as the class of 'segment vocoders' which give rise to a rate-distortion performance marked as region A in Fig. 1.2, characterized by significant reductions in bit-rate below that of 2.4 Kbps of LPC-10, with speech quality comparable to that of LPC-10.

While there are several well-established coders such as the MELP, MELPe and MELP suite [M08] that operate at the bit-rates of LPC-10, but with superior speech quality, the primary difference is that we focus on the class of segment vocoders, which continue to use linear-prediction framework, and differ primarily in the way the segmental units are defined, derived, modeled, quantized, duration normalized and synthesized.

This view of the class of segment quantization vocoders that try to realize ultra low bit-rate speech coding sees them as an evolution from the pivotal LPC-10 operating point, and with LPC-10 as the reference, leads to expecting ultra low bit-rate performances of comparable speech quality and bit-rates lower than 2.4 Kbps progressively down to 100 bps. While there would be considerable deviations from the essential LPC-10 framework, such as in the HMM based recognition-synthesis frameworks, or in the way the unit models are derived (e.g. phonetic vocdoer in [PD89] or by temporal decomposition, VQ, multigrams and HMM in [CBC98a]), the primary reference to LPC-10 serves more to underscore the advantage realized in moving from a scalar quantization or vector quantization of the linear prediction coefficients away towards quantizing them as variable length segments, regardless of the underlying feature representation, how they are derived by automatic segmentation, how they are modeled, and how they are eventually synthesized.

In this context, it is important to make two observations with regard to realizing better quality than that of the 2.4 Kbps LPC-10 standard in the segment vocoder framework, as shown in Fig 1.2, with the second arrow reaching the rate-distortion region marked B. This appears implausible at the outset, but is actually possible, by virtue of (a) better quantization of the LP coefficients, such as what certain classes of segment quantization can actually achieve (as pointed out in Sect. 1.3.5 to follow) and, (b) better (or even no) modeling and quantization (or even transmission) of the residual (as in the algorithms to be discussed in Chap. 6):

a. With regard to the LP parameter quantization, it is to be noted that there are a class of quantization techniques, all set in the vector quantization framework, that operate at significantly much higher bit-rates, such as 24 bits/frame with an effective bit-rate for spectral quantization as 1,200 bps (using a frame-rate of 50 frames/s) and above that are geared towards achieving high speech quality, referred to as 'transparent quality', characterized by the underlying spectral distortion of 1 dB [PA93, PK95]. Since the segment quantization techniques employed in segment vocoders operate at significantly lower bit-rates, at the outset, it seems natural to conclude that these two ranges of operation (and the corresponding quality) are not congruent. However, in line with the reasonings given in Sect. 1.3.5 and also based on the performance trends reported in several segment quantization techniques (e.g., [KCT91, LF93, B95, HR08]), which report reaching spectral distortions of 2.5–2 dB (and even less) at low bit-rates of 500 bps and less, it seems plausible that there could be a convergence of these two directions of LP quantization, i.e., the high-rate quantizers and the very low rate segment quantizers, and consequently, allow for targeting speech quality far superior to that of the LPC-10 baseline.

b. It is known that the quality of the resulting LP synthesis speech is strongly influenced (and limited) by the residual modeling employed, namely the categorical voicing/unvoicing decision and the corresponding excitation by a pulse train with an appropriate pitch or a random noise sequence. Much effort in realizing a more realistic model of the residual has resulted in coders such as the MELP [MB95, M08]. As a baseline, it can be noted that when the residual is retained as it is, i.e., without any modeling or quantization and used in the LP synthesis at the decoder, with only LP parameter quantization, such a baseline yields the best quality speech possible, under the constraint that only the LP parameters are quantized, and thereby truly reflects the LP quantization efficiency (note that, with no quantization of the LP parameters and the residual, the synthesized speech is identical to the original speech). Thus, techniques which attempt to do little or no processing or change of the residual can actually yield better quality speech at the decoder, than what is possible by the LPC-10 baseline system. Towards this, we will see that the no-residue transmission system proposed by us in [RH09, R12] and discussed in Chap. 6 can target subjective performances superior to that of LPC-10.

A more detailed treatment of the segment quantization principles and the various techniques that have been proposed and studied to date is given in Chap. 2, before proceeding to the subsequent chapters that focus on the unit-selection based segment quantization in detail. The perspective of how unit-selection segment quantization arises as a natural, though very distinctively different paradigm, from variable length segment quantization is now presented.

1.3 Clustered Codebook to Continuous Codebook

In this section, we see a set of apparently disparate, but yet converging observations that help us to reason about how a segment quantization paradigm which uses a segment codebook derived from a long continuous sequence of vectors (such as from 4 h of continuous speech) can lead to quantization performances, that were hitherto considered capable of being achieved only by high-rate vector quantization techniques.

1.3.1 Clustered Segment Codebook

While dealing with segment quantization, the segment codebook which is made of a set of (say, N) variable length segments plays a crucial role in determining the rate-distortion performance. As we will note in more detail in Chap. 2, the first attempt to design an optimal variable length segment codebook of a given size N was by Shiraki and Honda [SH88], who used the framework of segmental K-means algorithm to derive a clustered segment codebook that minimizes the average quantization distortion defined between an input segment and the nearest variable length segment in the codebook, the average being calculated over a large collection of naturally occurring input segments from continuous speech. A clustered codebook of size 1,024 (10 bits/segment) was shown to offer down to 2.5 dB spectral distortion in [SH88, HR06] (also in this book, Sect. 4.3.1). While the distortion would continue to decrease with increase in codebook size, no major efforts have been expended further in exploring how large a 'clustered' codebook size could be used to reach even lower spectral distortions, primarily owing to the complexity of the codebook design algorithms such as the segmental K-means algorithm.

1.3.2 Vector Quantization Performance Trends and Limits

In this context, it is worthwhile examining the fundamental results of quantizing linear prediction coefficient vectors using vector codebooks (i.e., vector quantization) of increasing size. Paliwal and Kleijn [PK95] showed the following:

(a) From actual measurements of average nearest-neighbor quantization error using randomized codebooks of sizes varying from 2 to 16,384, it was seen that a random codebook of size 14 bits/vector (i.e., size 16,384) can yield a spectral distortion upper bound of 2 dB, implying that an optimally designed codebook of same size will yield distortions less than 2 dB.
(b) By means of linear extrapolation of the above measured rate-distortion trend—involving the linear relationship between the log root-mean-squared distance in a uniformly distributed set of points (LPC vectors) and their density—it was concluded that a random codebook of size 20 bits/frame (1,024 K vectors or frames) can yield a spectral distortion upper bound of 1 dB, which is now well established as the important 'transparent quality' quantization [PK95].

The fact that a random vector codebook or a random segment codebook yields the above linear performance can be further interpreted as follows: As the codebook size is doubled progressively, the spectral distortion continues to reduce linearly. However, considering the vector quantization case of [PK95], reaching codebook sizes of 20 bits/vector (even if it be a random codebook) is formidable primarily from computational complexity considerations. This is precisely what led to the important developments in realizing transparent quality quantization (1 dB spectral distortion) with 24 bits/frame split vector codebook schemes [PK95], which allowed manageable complexities of the split codebooks.

However, even if we were able to double the (random) codebook sizes up to 20 bits/vector at a frame rate of 50 frames/s (using an effective frame size of 20 ms), this would demand an increase of bit-rate by only 50 bps for every doubling of the codebook, eventually leading to an overall bit-rate of only 1,000 bps (for a 20 bits/vector codebook) for the quantization of spectral information (LP coefficient vectors). This does fall within the low bit-rate range, considering that additional bits need to be spent in the quantization of the residual information in a LPC-10 framework. Clearly, this shows that the effective bit-rate needed for high quality spectral quantization is not a bottleneck by itself, as much as the computational complexity in handling such large codebooks.

1.3.3 Random Segment Codebooks

An interestingly similar result was obtained by Roucos, Schwartz and Makhoul [RSM83] for segment quantization using actual measurements on speech LP coefficient vectors (LARs), using both clustered codebooks (segment) and randomly selected segment codebooks of large sizes (up to 8,192 variable length segments or 13 bits/segment). In this work, [RSM83] showed that the mean-square quantization error (in dB) reduces linearly for every doubling of the codebook size (or directly with bits/segment) and that, for a given spectral distortion, the random codebook was only 2 bits/segment larger than the clustered codebook (i.e. 4 times larger), or conversely, for a given segment codebook size (e.g. in the upper end of 11–12 bits/segment), the random codebook had only 1–1.5 dB poorer mean-square quantization error than a carefully optimized clustered segment codebook.

1.3.4 Vector to Segment Quantization Performance Retention

Here, we consider a very important result by Svendsen [S94], in the context of moving from vector quantization to variable length segment quantization. This work establishes that it is possible to use a vector quantizer on segments larger than a single frame in the input speech, so as to be able to realize effective frame-rates lower than the vector based frame-rate and which in turn lowers the effective bit-rate by a factor of 2 under the constraint that the spectral distortion incurred in quantizing a segment instead of a vector by a given vector quantization codevector remains same, i.e., at the 1 dB transparent quality spectral distortion threshold used in the paper [S94].

This work poses the question whether a 'segment representation vector' (possibly also quantized) can be made to represent a variable length segment instead of a single input vector, with the segmentation being derived under constrained–optimization conditions, of limiting the resulting spectral distortion within a desired threshold (e.g. 1 dB). It uses a maximum likelihood formulation of deriving the optimal segmentation in such a manner that the resulting distortion $D(m)$ which is a function of only the m segments, and being the sum of the spectral distortion of the individual segments each obtained as the average spectral distortion of the vectors in the segment when represented and quantized by the segment representation vector. This essentially serves to reduce the frame-rate of the resulting segment representation and quantization, which is perhaps more aptly called the segment rate, since there are now less segments than single frames in 1 s of speech that are represented (and quantized) using the same segment representation vector (with either a simultaneous or a subsequent scalar or vector quantization).

Through this formulation, the work shows that a 40 bits/frame scalar quantizer of LP-cepstral parameters can be made to work at an effective segment rate that is approximately two times less than that of the original frame-rate, thereby leading to an effective bit-rate of 22.6 bits/frame, even while preserving the 1.06 dB spectral distortion. In the following section, we will see how this result contributes to the converging viewpoint that helps reason about how segment quantization can lead to spectral distortions close to that of limiting vector quantization with very large vector codebooks, even while incurring very low bit-rates.

1.3.5 A Converging Viewpoint

In this section, we put together the above stated fundamental trends (in Sects. 1.3.1, 1.3.2, 1.3.3 and 1.3.4) to carry out two types of reasoning to show the possibility that segment quantizers with reasonable size segment codebook sizes (or unit databases) can reach the low spectral distortions characteristic of the 1 dB transparent quality quantization (presently in the realm of high rate vector quantizers) at ultra low bit-rate ranges.

1.3.5.1 Reasoning I

In this reasoning, we put together the above stated facts that (i) a random vector codebook of 14 bits/vector can yield a spectral distortion of 2 dB and less and a random codebook of 20 bits/vector can yield a spectral distortion of 1 dB (transparent quality) and less, and (ii) a similar scenario arises with segment codebooks, wherein a random segment codebook of size 1,024 can yield 2.5 dB spectral distortion, with a linear trend in the decrease in spectral distortion for every doubling of the segment codebook, and that a random segment codebook is only 2 bits/segment larger than a clustered segment codebook in realizing comparable spectral quantization distortion. This reasoning is illustrated in Fig. 1.3, which is divided into two parts, as discussed below.

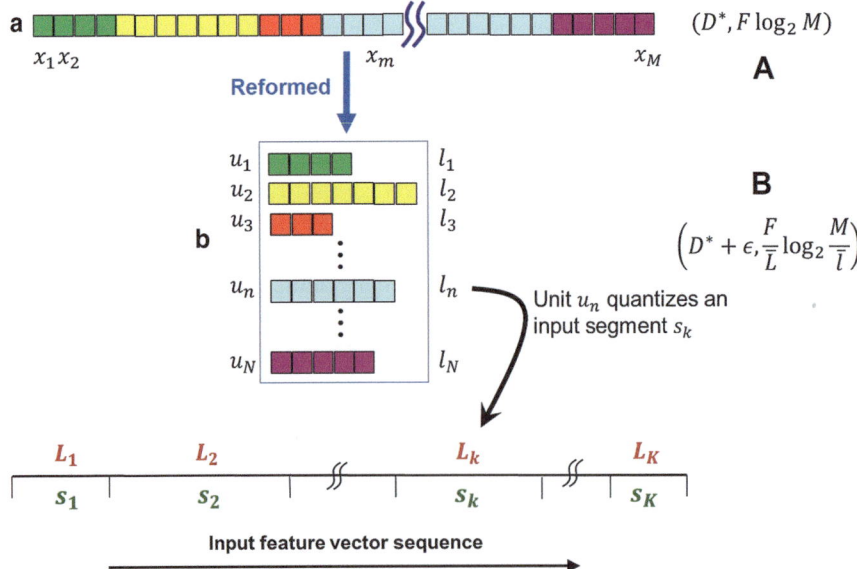

Fig. 1.3 Reasoning I: Vector and segment quantization scenarios and associated (Distortion, Rate)

1.3 Clustered Codebook to Continuous Codebook

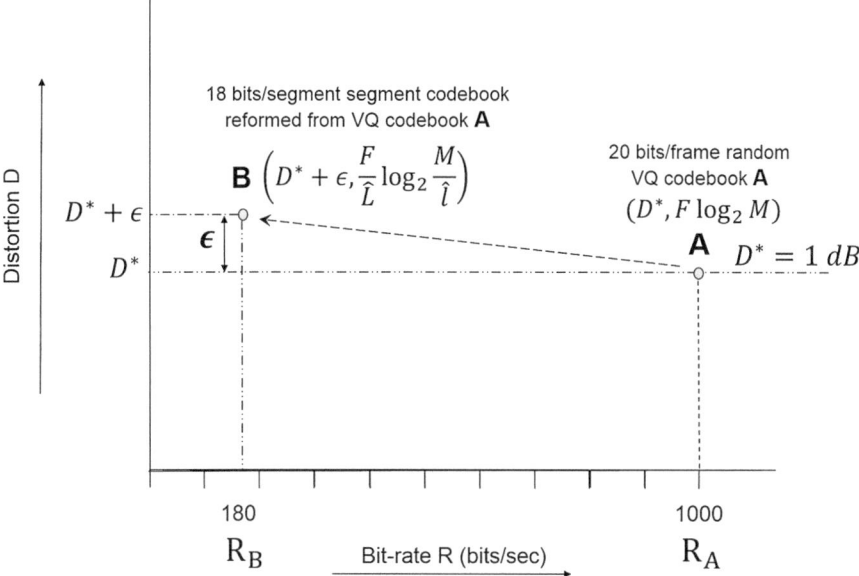

Fig. 1.4 Reasoning I: The quantization scenarios marked in the rate-distortion space

Part (a)

Consider a long continuous speech represented in terms of a sequence of LP coefficient vectors (e.g. LARs as in [RSM83]). Let the number of vectors in this continuous sequence be M. Typically, the above results showed that for $M = 2^{20} = 1024$ K vectors or 20 bits/vector rate, such a continuous codebook is a random codebook, but can reach $D^* = 1$ dB spectral distortion limits. This corresponds to about 4 h of continuous speech. The corresponding bit-rate is $R_A = F \log_2 M = 1,000$ bit/s (for $F=50$ frames/sec frame-rate). We denote this as Case A in Fig. 1.3 and marked as point A with (Distortion, Rate) of $(D^*, F \log_2 M)$ in Fig. 1.4.

Part (b)

Now consider segmenting this long vector codebook and 'reforming' it into a segment codebook U of N contiguous variable length segments (or units) $U = (u_n, n = 1, \ldots, N)$, with an average segment length $\hat{l} = \sum_{n=1}^{N} l_n$, where l_n is the length of the individual units $u_n, n = 1, \ldots, N$. Clearly, $N = M/\hat{l}$. This variable length segment codebook can now be used to perform segment quantization of an input speech of 1 s, with typical number of segments $K = 10$ being determined by the typical speaking rate of 10 phonemes/s, with each segment assumed to be a phoneme-like unit. The effective bit-rate of this segment quantization is

$R_B = K \log_2 N$ bps which is also obtained equivalently as $(F/\hat{L})\log_2 N = (F/\hat{L}) \log_2(M/\hat{l})$ bps, where F is the frame-rate which is 50 frames/s, for a frame size of 20 ms and $\hat{L}$ is the average length (in frames) of the input segments as derived by the segmentation resulting from segmentation and quantization of the input speech using the segment codebook U. Typical values of $K = 10$ correspond to $\hat{L} = 5$, i.e., the number of segments per second of speech is 10, with each segment being on an average 5 frame long (or 100 ms duration). Note that since the segment codebook U is formed by a similar segmentation of the original vector codebook (of size M vectors), the average length of the units in the segment codebook $\hat{l}$ is likely to be same as $\hat{L}$, thus allowing us to use $\hat{l} = \hat{L} = 5$.

Starting from the primary result (in Part (a) above) that a 20 bit vector codebook (4 h of speech) can reach 1 dB spectral distortion, we first note that a 18 bits/segment segment codebook derived by segmenting this long vector codebook is identical to the 20 bit codebook, with a variable length segment entry in the segment codebook being made of a short sequence of contiguous frames of the vector codebook. Thus, segment quantizing an input speech using the segment codebook can yield spectral distortions comparable to that obtained by vector quantizing the input speech by the long vector codebook, but conditioned by the following observation: When each LP coefficient vector in the input speech is quantized by the lowest distortion vector in the vector codebook, every vector in the vector codebook was a potential candidate for being the quantization vector of the input vector. However, in the segment quantization scenario, a sequence of vectors in the input speech gets quantized as a segment to the best segment in the segment codebook, and by this, a vector in the input speech segment is constrained to be quantized by one of the vectors in the same segment (unit) that quantizes that input speech segment. This is illustrated in Fig. 1.3 as Case B.

Due to this constraint, each vector in the input speech is not quantized by the least distortion 'vector' in the vector codebook, but by the vector in the least distortion 'segment' in the segment codebook, thereby resulting in a higher quantization distortion. But this increase in quantization distortion is bound to be marginal, considering that the segment selected from the segment codebook to quantize an input segment (a sequence of vectors) is indeed a segment that minimizes the segmental measure of quantization distortion, and hence each vector in the input segment does not suffer any unduly higher quantization distortion than it would have, had it been quantized by the best vector from the entire vector codebook in an unconstrained manner. Therefore, it is not unreasonable to expect the 18 bits/segment random segment codebook (obtained by reforming the original 20 bits/vector vector codebook) might actually reach the same quantization distortion D^* as did the 20 bits/vector vector codebook or marginally higher distortions $D^* + \epsilon$.

We denote this as the point B in Fig. 1.4, marked by a (Distortion, Rate) of $(D^* + \epsilon, (F/\hat{L})\log_2(M/\hat{l}))$. For typical values of $M = 2^{20}$, $\hat{l} = \hat{L} = 5$ and $K = 10$, we get $N = 2^{18}$ and $R_B = 18 \times 10 = 180$ bit/s. Note that the corresponding bit-rate in bits/vector is $(\log_2 M)/\hat{L} = 18/5 = 3.6$, which turns out to be the effective bit-rate for quantizing a vector of input speech at only marginally higher distortions of $D^* + \epsilon$, which is remarkably lower than the 20 bits/vector quantization limit that is considered necessary to reach the $D^* = 1$ dB transparent quality spectral distortions [PK95].

1.3.5.2 Reasoning II

An alternative and supporting evidence to the above line of reasoning emerges from the result of [S94] discussed in Sect. 1.3.4. This result showed that it is possible to use a vector quantizer (VQ) codevector to represent (and quantize) all the vectors in a variable length segment, when the segments are derived under optimality constraints (such as the maximum-likelihood segmentation used in [SS87, S94]), without increasing the spectral distortion (above a specified threshold, which in the work of [S94] was set to $D^* = 1$ dB). Using this result here, we provide the second reasoning, as illustrated in Fig. 1.5, in three parts, namely, (a) a code vector in a 20 bits/frame VQ codebook quantizes a vector in the input speech, (b) the same code vector quantizes a segment in the input feature vector sequence and (c) the VQ codebook is reformed into a variable length segment codebook, and a segment (unit) in the segment codebook quantizes the same input segment (as in part (b)), in the place of a codevector.

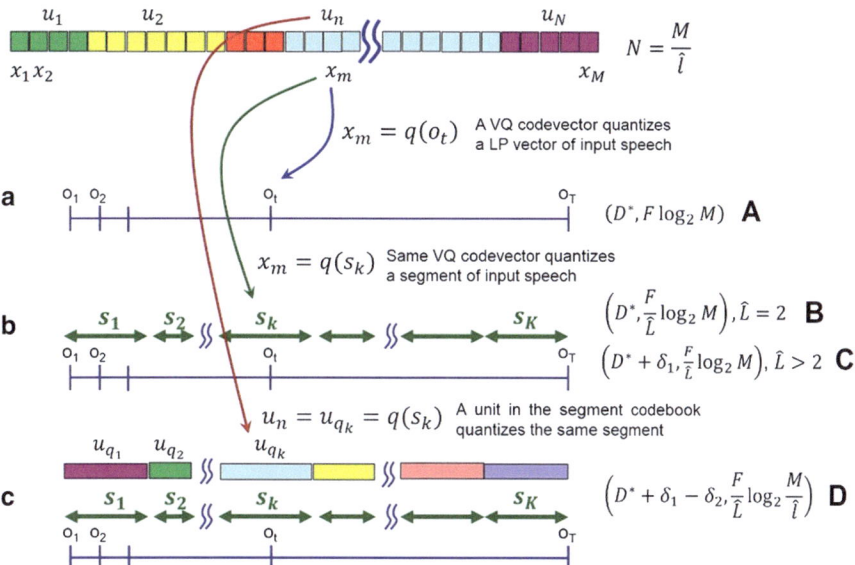

Fig. 1.5 Reasoning II: Vector and segment quantization scenarios and associated (Distortion, Rate)

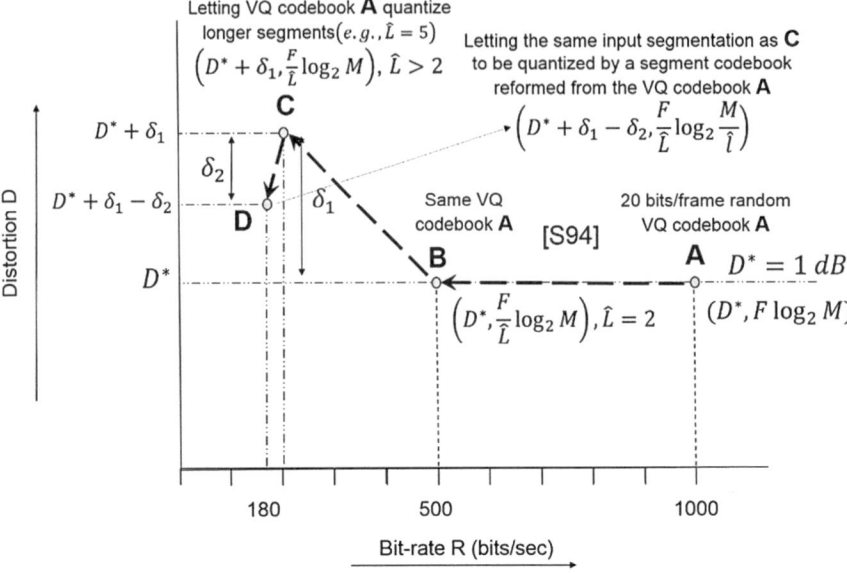

Fig. 1.6 Reasoning II: The quantization scenarios marked in the rate-distortion space

Part (a)

Let the vector quantizer with the long 20 bits/frame VQ codebook (of size $M = 2^{20}$ vectors) realize a certain spectral distortion, say $D^* = 1\ dB$, as considered in Case A of Reasoning I. This is shown as Case A in Fig. 1.5 with the baseline (Distortion, Rate) of $(D^*, F\log_2 M)$ bps, as also marked in the rate-distortion space in Fig. 1.6 as the point A.

Part (b)

With reference to part (b) of Fig. 1.5, we consider the result of [S94] as indicated above, by which it is possible to let each of the codevector in this M size VQ codebook to quantize a sequence of vectors making up a variable length segment in the input speech, without increasing the overall distortion. If the average segment length (in the resulting input speech segmentation) is $\hat{L}$, then the effective bit-rate is $(F/\hat{L})\log_2 M$ bps, with $(F/\hat{L})$ now representing the reduced segment rate. For $\hat{L} = 2$, we get a remarkable halving of the frame-rate to 25 segments/s, (though shown in [S94] for the higher rate scalar quantization of 40 bits/frame, reducing it to 22.6 bits/frame) while retaining the 1 dB spectral distortion. This is shown as Case B in Fig. 1.5, with the associated (Distortion, Rate) of $(D^*, (F/\hat{L})\log_2 M)$, with $\hat{L} = 2$. This is also marked as a transition from Case A to Case B in the rate-distortion space in Fig. 1.6, with associated bit-rate of 500 bps (half of the 1,000 bps of Case A) and at the same distortion D^*.

1.3 Clustered Codebook to Continuous Codebook

It is now easily conceivable that we can increase $\hat{L}$ to as much as 5, leading to the natural segment rates of 10 segments/s and an effective bit-rate of 200 bps. Values of $\hat{L} = 5$ arises from considerations of quasi-stationarity of speech, where it can be assumed that most phonemes have natural durations of that order, characterized by steady-state regions and transitions from and to their left- and right-context phones, but with spectral variability within a segment limited to the extent of allowing the entire segment to be quantized by a single vector (say x_m, as shown in Fig. 1.5) from the 20 bits/vector VQ codebook. This is shown as Case C, with the associated (Distortion, Rate) of $(D^* + \delta_1, (F/\hat{L})\log_2 M)$ for $\hat{L} > 2$ and marked as C in the rate-distortion space in Fig. 1.6, with a transition from B to C associated with a decrease in bit-rate from 500 bps to 200 bps (for $\hat{L} = 5$) and an increase in distortion to $D^* + \delta_1$.

Part (c)

As part (c) in Fig. 1.5, we extend the reasoning further and let a segment (say, unit u_n as shown in Fig. 1.5 under Case D) centered at the vector (x_m) in question in the 20 bits/vector VQ codebook to quantize the input segment that was originally quantized only by vector x_m. By this, the spectral distortion is clearly decreased, with reference to that achieved by Case C. Use of a segment codebook, formed from variable length segments of this type, therefore serves to reduce both the bit-rate and spectral distortion further to the scenario outlined above: (i) reduction in bit-rate is realized as the number of bits used to represent a segment (or unit) in the segment codebook is reduced from 20 bits/frame to $\log_2(M/\hat{l}) = 20 - \log_2 \hat{l} = 18$ bits/unit, where $\hat{l}$ is the average segment length in the segment codebook (using typical values of $\hat{l} = 4$ or 5, as outlined in Part (b) of Reasoning I), formed by such segments from the 20 bits/frame vector codebook, and (ii) reduction in spectral distortion is realized (with respect to the distortion of Case C), by having requisitioned an entire segment (from the segment codebook) to quantize an entire input segment, where originally, only a single vector from the 20 bits/frame vector codebook was used. This is shown as Case D in Fig. 1.5, with the associated (Distortion, Rate) of $(D^* + \delta_1 - \delta_2, (F/\hat{L})\log_2(M/\hat{l}))$, and marked in the rate-distortion space in Fig. 1.6, with a rate of 180 bps, and a transition from C to D.

Once the above two lines of reasoning are established, the accompanying arithmetic shows that a random variable-length segment codebook of size 18 bits/segment, reformed from a 20 bit/vector random vector codebook can realize spectral distortions close to 1 dB (i.e., $D^* + \delta_1 - \delta_2$), at an equivalent 3.6 bits/frame given an average segment (unit) length of 5 frames in the variable length segment codebook (or unit database), resulting in an overall bit-rate of 180 bits/frame (at a frame-rate of 50 frames/s). It is easy to extend this result further to show that longer the average length of the segment in the segment codebook (i.e., the unit of quantization gets longer, with higher $\hat{l}$), lower is the overall bit-rate, and at the same time better the subjective quality of coded speech, considering that longer segments in the input speech are now approximated by longer naturally occurring segments in the segment codebook (i.e., higher $\hat{l}$ leading to higher $\hat{L}$), which help retain the intra-segment co-articulatory effects, even if the longer segment serves as a constraint for a sequence of vectors in the input segment to have to be quantized by only those vectors in the quantizing segment (unit) from the segment codebook (unit database).

This also provides a means to have very large segment codebook sizes and sets the basis for a new paradigm to emerge—that of moving away from the conventional segment quantization as finding the best quantizing segment from a segment codebook for a given input segment—to the notion of unit-selection based segment quantization using a large unit-database (such as 'reformed' from continuous speech data, as in the reasonings above), as in text-to-speech synthesis by unit-selection based concatenative synthesis, as will be discussed further in the next section.

1.4 Speech-to-Speech Synthesis by Unit-Selection

As seen in the previous section, the shift from using small clustered segment codebooks to very large continuous codebooks leads to the possibility of segment quantization within the unit-selection framework, as is now well established in text-to-synthesis systems. Even without dwelling into details of what such a unit-selection mechanism entails, it is easy to visualize how the entire chain of TTS gets adapted to what can be called 'speech-to-speech' (STS) synthesis, i.e. driving the synthesis chain with abstract features derived from the input speech signal to be quantized, rather than the phonetic and prosodic units derived from text, and continuing to leverage the subsequent unit-selection mechanism using a large unit-database of units that now replace the conventional segment codebook in a segment quantization system.

Figure 1.7 shows the schematic of a typical unit-selection based TTS system. Such a system starts from the input text, which is converted to a sequence of phonetic units by means of letter-to-sound rules (also called grapheme-to-phoneme or G2P convertor), which are referred to as the target $t_1, t_2, \ldots, t_k, \ldots, t_K$. The target $t_k, k = 1, \ldots, K$ for a text defines the string of phonemes required to synthesize the text, and is primarily made of the phonetic identity as well as the prosodic content of that phoneme such as duration, pitch and intensity. In a more complete setting, the phonetic aspect of t_i can be in the form of a multi-dimensional feature vector representing various aspects of preceding and following phonetic context in the text, comprising a set of distinctive features such as vowel vs. consonant, voicing, consonant type, point of articulation and vowel height, length and rounding [HB96].

1.4 Speech-to-Speech Synthesis by Unit-Selection

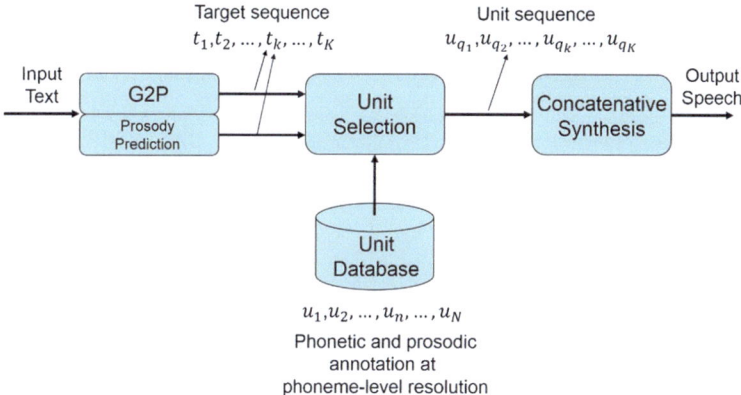

Fig. 1.7 Generic text-to-speech (TTS) system based on unit-selection and concatenative synthesis

The role of unit-selection in a TTS system is to select a set of units $u_{q_1}, u_{q_2}, \ldots, u_{q_k}, \ldots, u_{q_K}$ that are a closest approximation of the given target sequence $t_1, t_2, \ldots, t_k, \ldots, t_K$ in such a manner that the unit-sequence $u_{q_k}, k = 1, \ldots, K$, when synthesized by means of a concatenative synthesis module, realizes the speech corresponding to the input text, firstly, by being faithful to the phonetic content, secondly, by having the correct prosodic variation that makes it sound natural, and thirdly, by being of acceptable quality without artefacts arising from the concatenation of the units $u_{q_k}, k = 1, \ldots, K$ drawn from diverse contexts in the unit-database or from the signal processing that converts the units to the speech signal.

Current state-of-the-art TTS synthesis systems based on the above broad principle, are able to achieve phenomenally high quality of the synthesized speech, for most languages around the world, due to highly evolved techniques that go into almost all the components that make up the system— the G2P, context specification, prosodic prediction from text, large unit databases (anywhere from 2 to 10 h of speech), annotations of the unit-database that characterize the various phonetic, contextual and prosodic aspects, highly refined unit-selection procedures with well-defined unit-costs and joint-costs, and sophisticated signal processing in the actual synthesis of speech from the units retrieved from the unit-database.

Imagine now a speech-to-speech synthesis (STS) system that works with the target $t_1, t_2, \ldots, t_k, \ldots, t_K$ being derived not from an input text but from the input speech signal. It is likely that the target can be derived more precisely from the speech signal than from the text, given that this calls for signal processing to derive the phonemes, their context and prosodic aspects. However, two important variations are needed to mark such a shift from TTS to STS: (i) the target are specified as spectral feature vectors, such as the linear prediction coefficient vectors or MFCCs given as $o_1, o_2, \ldots, o_t, \ldots, o_T$, i.e., T feature vectors (corresponding to the K phoneme targets for the same underlying textual content, but now with $T \gg K$, since each phone label would correspond to several, (e.g., 10), frames). This excludes dependence on higher level phoneme recognition kind of transcription

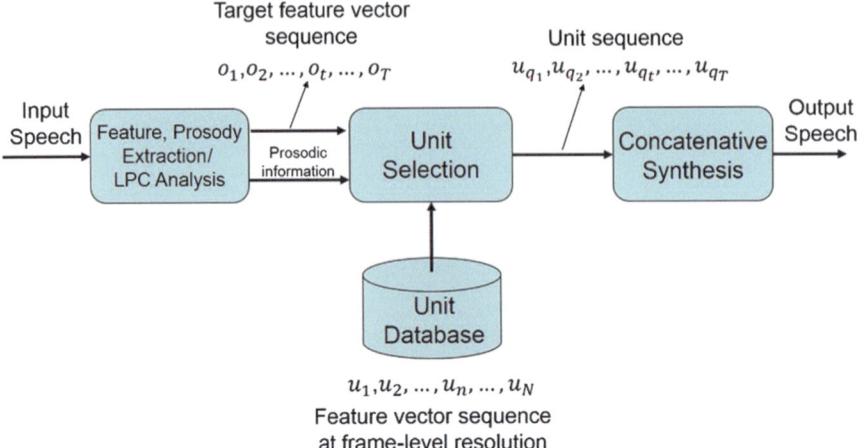

Fig. 1.8 Generic speech-to-speech synthesis set in the unit-selection based concatenative synthesis, as a parallel to TTS

which would come with their associated errors, and allows the target to represent phonetic content of the input speech in the form of acoustic feature vectors, which now represents a fine-grained representation of the input speech signal, (ii) The unit database can now be a long continuous sequence of feature vectors (the same as the target, e.g. LP vectors or MFCC) with prosodic annotation such as pitch, duration and intensity. This is illustrated in Fig. 1.8.

The role of unit-selection in a STS system is to select a set of units $u_{q_1}, u_{q_2}, \ldots,$ $u_{q_t}, \ldots u_{q_T}$ that are a closest approximation of the given target sequence $o_1, o_2, \ldots,$ $o_t, \ldots, o_T$ in such a manner that the unit-sequence $u_{q_t}, t = 1, \ldots, T$, when synthesized by means of a concatenative synthesis module, realizes an output speech corresponding to the input speech.

It can be seen that this is in the lines of copy-synthesis, where a synthesizer (using any of the many possible synthesis models, e.g. LP synthesis, formant synthesis, sinusoidal modeling, HNM etc.), is driven by parameters extracted from an input signal, and how close the synthesized speech is to the input signal is a useful measure of how good the synthesis model is, in the first place, and how good the front-end speech analysis is that extracts the various synthesis parameters needed to drive the synthesizer model [D97]. In the context of STS by unit-selection, as discussed here, the synthesis method is unit-selection based concatenative synthesis, and the synthetic speech is close to the input speech depending on the effectiveness of the feature vector measurement from the input speech and the unit-selection method in selecting units that ensure high fidelity to the spectral and prosodic target specification and low artifactual distortion in the concatenative synthesis and the synthesis itself.

It is clear that the fine-grained target specification (in terms of spectral and prosodic measurements) as extracted from the input speech signal are more accurate representations of the speech to be synthesized, in comparison to the course-grained (and error prone) target specification (of phonetic labels and prosodic prediction) from input text. As a consequence, the resulting synthesized speech can be as good or even better than the speech synthesized from text, considering the following considerations: (i) the STS system is driven by fine-grained acoustic feature vector target representation, which can be matched finely (in terms of spectral match) by the unit-selection procedure, (ii) the potential misrepresentation of input text through faulty G2P or prosodic prediction from text is completely avoided, as these are now derived directly from the speech signal by signal processing, (iii) the synthesis of speech from the unit-database represented in form of fine-grained spectral features is potentially of higher quality, such as is already in practice in the unit-selection based speech coding systems (such as using HNM or LPC synthesis).

With the above considerations in place, what needs to be stressed here is that current TTS systems in most languages have very high quality (subjective listening scores such as MOS of 3.5 and above) as typified in applications such as car-navigation, telephony information retrieval systems, iPhone SIRI, audio books etc.. Such speech is undoubtedly speaker dependent, as they work with a single speaker unit-database, but the resultant speech is of such high quality that it would definitely be rated as significantly better than any of the low rate coding standards operating in the range of 2.4 Kbps and less.

This then sets the basis for realizing as good or better quality of coded speech in a STS framework based on unit-selection principles. A major part of this book will indeed focus on such unit-selection frameworks (Chaps. 3–6), starting with the pioneering work of Lee and Cox [LC01, LC02], followed by our own contributions [RH06, RH07, HR08, RH08, RH09, R12] that are in the form of

i. unified, optimal and generalized unit-selection algorithms [RH06], [RH07]
ii. comparative study with vector quantization, matrix quantization and variable length segment quantization [HR08]
iii. generalizations over the Lee and Cox algorithms to render it more optimal [RH08] and,
iv. an interesting framework of joint spectral-residual quantization in unit-selection framework, that obviates the need to transmit any information about the residual [RH09], [R12]

1.5 Alternate Perspectives for Ultra Low Bit-Rate Speech Coding

Interestingly, the success of speech coding as a viable every-day technology probably rests also in its not having to reach up to a parallel 'human' performance benchmark or reference, as is the case with the other speech technologies, such as for example, the high human performance levels in speech recognition, understanding, speaker-, language- and accent-recognition, binaural hearing, source localization and source separation, etc., in contrast to the comparatively poorer performance levels of machines in the respective tasks, particularly in unrestricted domains, characterized by various dimensions such as large vocabulary, spontaneous and conversational speech, language perplexity, additive and convolutive noise, sound mixtures, etc.

The reason that speech coding seems to have no 'human' parallel is largely due to the problem definition that it is required to perform a lossy data compression of an original speech signal, in such a manner to progressively reduce the bit-rate required for its representation, transmission and reproduction over digital communication channels, even while ensuring a graceful (or even no) degradation of the quality of compressed speech, considered acceptable for human-to-human communication, such as in telephony, broadcast, storage, etc.

Human speech communication in its biological form has practically no equivalent of this process, except possible in two extreme and rather esoteric forms of communication modalities: One being the word of mouth communication, where speech heard by one person is memorized and spoken out by that person to another person, and the other being the invocation of the written form, wherein the heard speech is committed to a script, that can be further read out in spoken form. Undoubtedly, humans excel in both these two process, bringing in a variety of deep cognitive processes to play, starting from speech recognition, understanding, speaker and accent recognition, along with the various aspects of prosodic, expressive and emotional content of the heard speech, memorizing, and being able to reproduce the heard speech in a kind of imitation (or mimicry), restoring back, as far as possible, the various aspects of speech content, prosody, expressiveness and emotion.

In the second scenario alluded above, i.e., when the interim representation is not memory, but a form of script, here again the success of reproduction (in terms of how close the reproduced speech is to the heard speech) is largely governed by how well the heard speech in its totality (with respect to the phonetic and prosodic content, expressiveness and emotional aspects) is captured uniquely in the script form, how well it is deciphered when read back, and how well it is faithfully reproduced while speaking out. In both the above scenarios, the internal representation in memorizing the heard speech or the scriptural form of representation, corresponds to the coded form of speech, which allows faithful reproduction (such as at the receiver or decoder in a communication system), and the relative compression ratios possible with such representations would possibly set the achievable lower bounds for bit-rates and quality in speech coding. This is remarkably difficult to quantify, considering very little is still known as to wherein lies the speaker identity, language identity or accent identity and expressive and emotional content in the speech signal, despite significant progress in the understanding of the phonetic and prosodic content of a given speech signal and significant efforts in unraveling the speaker-specific information and recent efforts in similarly understanding expressive and emotional content of speech signals [NA05]; further difficulty lies in knowing, let alone quantifying, the way human cognition retrieves these multiple interwoven layers of information from the speech signal, and how these are represented in human memory and further retrieved and used for reproduction.

While the preceding paragraph may appear a bit too qualitative and venturing into a rather nebulous and as yet, poorly understood, cognitive dimensions of speech and language, a careful thought along these lines reveals the human performance reference that a speech coding system does indeed have. As we shall see further, such a setting can lend itself to asking how a computational framework for speech coding can be evolved that draws from the above qualitative chain of hearing-(script)-speaking, towards realizing effective speech coding systems. This could well be the underlying basis, though within practically well-defined analysis-by-synthesis formulations, for articulatory parameter estimation and articulatory synthesis based very low bit-rate speech coding in the early works of Flanagan et al. [FIS80] and later work on the articulatory parameter estimation problem (e.g. Parthasarathy and Coker [PC92] and Sondhi and Sinder [SS05]) for synthesis. These frameworks, which by and large represent remarkable, and perhaps the only very few, ventures into articulatory basis for speech coding might as well hold the promise of potential directions towards reaching truly ultra low bit-rate coders being able to operate at parsimoniously low representations of the speech signal, even while being able to deliver high quality natural sounding speech.

1.6 Applications of Ultra Low Bit-Rate Speech Coding

The applications of speech coding in general are well known, that of being able to transmit or store speech at low bit-rates, as constrained by a given physical communication (or storage) system or channel, even while being able to maintain the highest possible speech quality of the coded speech. While the broad categories of bit-rate and quality that most speech coders fall in were indicated in the early part of this chapter, we alluded to the ultra low bit-rate as posing special challenges and which have to be addressed at before realizing coders at par with those at higher bit-rates and which have reached maturity of becoming standards. It would therefore be appropriate to ask as to what the special applications of such a range of coders are, given the intrinsic challenges and yet to be realized viable operational status.

At the outset, the answer to such a question is that the challenge of reaching the lower limits of speech coding (as outlined in Sects. 1.1–1.5 in this chapter) is appealing and enticing simply as a scientific pursuit, and which is likely to unravel various new paradigms that would become effective to realize such coders, such as the example quoted in Sect. 1.5 about the articulatory basis for low bit-rate coding or even the more abstract notions hazarded in the same section. More similar frameworks and paradigms are bound to emerge in such a pure pursuit of the science behind such ultra low bit-rate realizations, and which, by virtue of bringing about paradigmatically promising and yet unexplored techniques, has the potential to have a bearing on the very nature of speech coding, even possibly to the extent of being disruptive enough to make such newfound high quality ultra low bit-rate coding principles offer qualities comparable to higher rate coders and thereby push the rate-distortion performances to new limits.

On a less ambitious note, the next answer to the question of why such ultra low bit-rates are needed is the dictum that 'lower the bit-rate, always the better', provided it is possible to ensure acceptable quality coded speech, which in turn can be qualified as at least intelligible and natural sounding, at such ultra low bit-rates. This is simply in keeping with the practical consideration that more speech channels can then be made to operate within a specified and available channel bandwidth.

Rather than be able to give only such a broad desideratum, it is in fact possible to identify specific niche applications where ultra low bit-rate speech coding is indeed of much use. One such niche is in HF communications (e.g. HF-ECCM—Electronic Counter Counter Measure [MLG95]) in defense applications, where it is required to build a ciphered and highly protected digitized speech communications on HF links. In such applications, even a baseline 2,400 bps coder (e.g. LPC10 or LPC10e) is considered to require too large a bit rate along with error protecting schemes, for this kind of channel. This immediately calls for the baseline coder to operate at much lower bit-rates (e.g. 800 bps, as identified in the NATO STANAG 4479 standardization efforts [MLG95]), so that the overall bit-rate including EDAC (Error Detection and Correction Code) is within the limits of 2,000–3,000 bps for such channels. In general, such a very low data rate (VLDR) deployment scenario typically arises in secure voice communications over narrowband channels and low-probability-of-intercept (LPI) communication and narrowband integrated voice/data systems [KF85, F93, W91]. Another application, also in communications setting, is the need for low bit-rate coders for underwater acoustic speech communications, considering the very low bandwidths available for such channels. This includes both ultra low bit-rate ranges, e.g. [Li11], at 600 bps in [DFGSL13] and at 300 bps in [Ji12] with frequency hopping communication system requirements with low probability of intercept and anti-interference, as well as higher rates as in [Luo11] at 2–2.4 Kbps and [GTLL05] at 5.4 Kbps for the speech encoder.

On a different note, the other niche application is in storage, where large speech data needs to be stored efficiently with as low a footprint as possible, even while being able to retrieve the same to yield high quality speech. Examples of audio storage abound already in multi-media applications, with various standards in place (e.g. the now popular and ubiquitous MP3 for general audio, with typical compression ratios of 1:11). However, when dealing with much higher compression ratios for speech, ultra low bit-rate speech coding would come into play typically to handle speech data as in audio books (also called talking book) or large archival data, as constituting the motivations in [BD84, VN97, LC01]. Specifically, [BD84] and [VN97] explored the rather esoteric application where the text transcript of the large archival type of speech is available. However, in a more general setting, it is clear that storage of industry grade bulk of speech (as in broadcast, archiving, multi-media, lectures, audio-visual content, telephony surveillance data, meeting capture data etc.) will call for as low rates as possible, even while ensuring that the speech retrieved satisfies certain baseline intelligibility and natural sounding requirements.

The desired quality of coded speech (i.e., the speech as would be retrieved from storage for later consumption) largely depends on the further use the coded and stored speech is put to, such as for audio-search and retrieval (as in lecture browsing or meeting capture data skimming) or forensic search (as in telephony surveillance data) or content based retrieval in general, for a wide range of analytics for actionable intelligence that is beginning to emerge as an important and specialized domain of application across a wide variety of data.

An important consideration in these two applications of ultra low bit-rate speech coding are various performance dimensions, in addition to the primary rate-distortion performance, such as coding delay, computational complexity, storage all impacting such coders' eventual realization in an embedded form for real-time operation or deployment on a server (e.g. cloud) for multi-channel real-time operations or off-line processing. It should however be noted that while the former application of communication scenarios require real-time operations with low delay and algorithmic footprint, the latter application of storage does not call for similar real-time processing requirement, as this can be done off-line, in the sense of allowing non real-time processing, where large data is compressed and stored at periodic intervals (say, end of day in a broadcast scenario or telephony surveillance scenario) for later retrieval.

1.7 Organization of the Book

While there are excellent review articles and collections of papers on speech coding techniques, standards and systems at bit-rates of 2.4 Kbps and above [F79, KP95b, K05, M08, MSH08], the range of 1 Kbps and below has received far less attention in the speech coding community. For instance, there are about 90+ essential papers in ultra low-bit rate coding over the last 3 decades, including the segment vocoder framework employing segment quantization principles, but there has been no effort so far in providing a consolidated framework of the techniques that operate at these ultra low bit-rates, except perhaps the early reviews of [SR83] and [JF94]. This book attempts to fill this gap and provides an overview of the various coding techniques proposed and developed till date for ultra low bit-rate coding, specifically in the framework of segment vocoders using segment quantization, in Chap. 2 while also offering a unifying perspective of the underlying operating principles that the various techniques share.

Chapter 3 introduces the pioneering work of Lee and Cox [LC01] in establishing the unit-selection framework for speech coding, and presents the essential algorithms for single-frame unit-selection and the segmental unit-selection proposed in their original papers [LC01, LC02]. Noting the dichotomy of these two methods, we also examine and bring out the inherent sub-optimality of the segmental unit-selection algorithm proposed by them.

Chapter 4 is exclusively devoted to our initial work in the unit-selection based coding framework [RH06, RH07] wherein we present optimal and unified unit-selection algorithms based on the one-pass dynamic programming framework and study its relative rate-distortion advantages over the sub-optimal segmental version of Lee and Cox [LC02]. Further in this chapter, we also provide a comprehensive comparative study of how the unit-selection framework fares in terms of rate-distortion performance over the clustered codebook based vector quantization, matrix quantization and variable-length segment quantization frameworks [HR08].

Following this, in Chap. 5, we examine the optimality and complexity tradeoff inherent in the segmental algorithm proposed by Lee and Cox [LC02] and our optimal one-pass DP formulation [RH08]. We propose and study a n-best lattice based search framework for the segmental version of Lee and Cox [LC02] unit-selection algorithm, and show how this enhances its optimality towards that of the one-pass DP algorithm, even while retaining the low complexity that the Lee and Cox segmental version inherently enjoys [RH08].

Chapter 6 provides an interesting and important contribution in the form of a joint spectral-residual quantization paradigm [RH09, R12], within the unit-selection frameworks of Lee and Cox [LC02] and our one-pass DP framework [RH06], [RH07], which obviates the need to transmit any information about the residual in a LPC vocoding framework. This result is likely to redefine the way LPC vocoding is done, wherein a speech-to-speech synthesis (STS) framework becomes completely self-contained in finding units that can be used for speech synthesis at the decoder without needing any information about the residual.

Chapter 2
Ultra Low Bit-Rate Coders

In this chapter, we present the definition and principles of ultra-low bit-rate coders. Here the emphasis is to point to the fact that this class of coders is typically the 'vocoders', which are 'parametric' coders that are essentially linear-prediction (LP) based vocoders. This is in contrast to the 'waveform' coders, which operate at the higher bit-rates. Among the various frameworks employed for realizing ultra low bit-rate speech coding, we restrict our attention to the generic linear-prediction (LP) based vocoding, (of which the LPC-10 coder [T82] sets the baseline) mainly for its rather ubiquitous adoption to a range of segment quantization vocoders as well as for its simplicity and effectiveness in establishing the low bit-rate operation and quality of coded speech. Here, the parameterization is based on the linear prediction filter model of human vocal tract wherein the linear prediction parameters (such as reflection coefficients, log-area-rations, line-spectral-frequencies, etc.) characterize the spectral shape of every frame of speech (typically of 20 ms duration) and the prediction error (residual) characterizes the excitation source of the speech production model. Coders have evolved to efficiently model and quantize both these information, i.e., the linear prediction parameters and the residual of each frame of the speech signal. However, by and large, the ultra low bit-rate coders have focused mainly on how to efficiently quantize the linear-prediction parameters either one frame at a time or by taking them as a sequence of frames together in the form of a 'segment'.

The main part of the book will focus on the various quantization schemes that have evolved towards quantizing the spectral parameters at the target bit-rates (of ultra low bit rate speech coding at less than few hundreds of bits/second). This chapter is organized to cover the various techniques in segment quantization, both conceptually and chronologically, as can be summarized under the following broad conceptual categories.

1. Vector quantization (VQ) and Matrix quantization (MQ) e.g., the VQ-LPC and MQ-LPC systems at 300–800 bps, viewed as fixed length segment vocoders.
2. Structure of generic variable length segment vocoders comprising various components and steps, such as feature extraction (e.g. LP analysis), automatic

segmentation, segment quantization (or joint segmentation and quantization), segment codebooks, duration modification, residual (or prosody) modeling, parameterization and quantization and synthesis. We provide a brief review of the various salient techniques proposed till date for each of these components and steps. Some of the techniques outlined include spectral transition measures, maximum likelihood segmentation, temporal decomposition for automatic segmentation, variable length segment quantization (VLSQ) using 2-level dynamic programming (DP) algorithm, joint segmentation and clustering algorithm (or the segmental K-means algorithm) for template segment codebook design, phoneme recognition and phonetic vocoders, template and HMM codebooks, space sampling for segment distortion and duration modification, pitch profile quantization, etc.

3. Rate-distortion constrained variable length segmentation and quantization, also referred to as R/D optimal linear prediction.
4. HMM based recognition-synthesis paradigms (sub-300 bps to 1 Kbps) focusing on HMM based phone class modeling and phone recognition in the HMM framework, followed by HTS based synthesis at the decoder.
5. ALISP units and refinements, focusing on the unit definition and modeling, segmentation and synthesis, with variants including definition of long synthesis units and short synthesis units by dynamic unit selection in a corpus based approach.
6. Unit selection paradigms, that mark a major departure from the notion of clustered codebooks or HMM codebooks to use of long continuous codebooks, in the form of single-frame vector codebooks or variable length segmental codebook, and which are used for segment quantization by unit-selection principles, as derived from TTS techniques.

On a related note, with regard to the LP parameter quantization, it is to be noted that we shall not attempt to review another important, and what can possibly be considered as mainstream class of quantization techniques, all set in the vector quantization framework (of quantizing vectors) and that operate at significantly much higher bit-rates, such as 24 bits/frame and with an effective bit-rate for 'spectral quantization' of 1,200 bps (using a frame-rate of 50 frames/s) and above. These are geared towards achieving high speech quality, referred to as 'transparent quality', characterized by the underlying spectral distortion of 1 dB [PK95]. In contrast, the focus here is exclusively on 'segment quantization', typically set in a segment vocoder framework, and operating at the lowest end of the bit-rate range of few hundreds of bits/second and less. In this regard, in addition to such high-rate vector quantizers (details of which can be found in the very comprehensive treatment of LP parameter quantization [PK95]), we also exclude from our consideration, another class of quantizers, also operating in the same bit-rate ranges, but based on exploiting inter-frame correlation between LP parameter vectors (LSF) such as [D04], [G07], [DP10].

2.1 Vector and Matrix Quantization

Figure 2.1 shows the basic LPC segment vocoder framework. Here, the speech signal is first analyzed at typical frame size 20 ms (in [T82], the standard LPC-10 vocoder uses a 22.5 ms frame size) yielding a LP parametric vector of dimension $p = 10$ corresponding to a LP order 10 in the LP analysis. Each frame also yields a residual associated with the LP parametric vector, which is further processed to derive the voiced/unvoiced decision, pitch and gain, which are the excitation parameters used to resynthesize speech at the decoder. LPC-10 quantizes the parametric vector (reflection coefficients) using non-uniform scalar quantization for each of the coefficients as described in [T82] with 41 bits/vector (and as shown in Fig. 2.2) and the excitation parameters (voicing, pitch, gain, sync) by scalar quantization with 13 bits/vector, yielding a total bit-rate of 54 bits/frame or 2.4 Kbps for a frame-rate of 44.4 frames/s. The quality of the LPC-10 coder is typically expressed as 'synthetic' quality, given in terms of subjective quality MOS score of 2.3.

As noted in Sect. 1.2, LPC-10 sets the basis for a class of low and ultra low bit-rate speech coding, all typically in the same LP vocoder framework, but essentially differing in the way the LP parameters are quantized using various techniques such as vector quantization, matrix quantization, variable length segment quantization, etc., progressively striving to reduce the effective bit-rate down from 2.4 Kbps, but keeping the LPC-10 quality as the target to achieve.

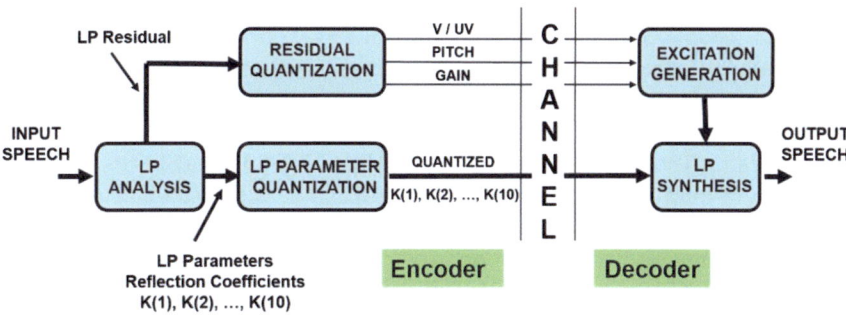

Fig. 2.1 Generic structure of the LPC-10 vocoder

	Pitch/Voicing	Gain	Sync	K(1)	K(2)	K(3)	K(4)	K(5)	K(6)	K(7)	K(8)	K(9)	K(10)	Total	Error Correction
VOICED	7	5	1	5	5	5	5	4	4	4	4	3	2	54	0
UNVOICED	7	5	1	5	5	5	5							33	20

Fig. 2.2 Bit-rate allocation for quantization of reflection coefficients, pitch and gain in LPC-10 vocoder

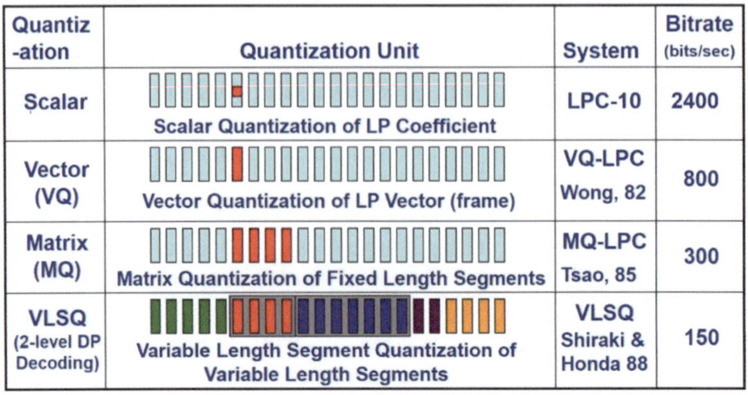

Quantiz-ation	Quantization Unit	System	Bitrate (bits/sec)
Scalar	Scalar Quantization of LP Coefficient	LPC-10	2400
Vector (VQ)	Vector Quantization of LP Vector (frame)	VQ-LPC Wong, 82	800
Matrix (MQ)	Matrix Quantization of Fixed Length Segments	MQ-LPC Tsao, 85	300
VLSQ (2-level DP Decoding)	Variable Length Segment Quantization of Variable Length Segments	VLSQ Shiraki & Honda 88	150

Fig. 2.3 Schematic of the various quantization schemes for LP parameter quantization and corresponding bit-rates

Figure 2.3 shows a schematic of the manner in which the LP parameter vector is quantized by the various quantization techniques referred above, namely, scalar quantization, vector quantization, matrix quantization and variable length segment quantization. Figure 2.4 shows a schematic of the feature space of the LP parameter vectors, where the input LP parameter vector sequence is quantized differently by a vector quantizer, matrix quantizer and variable length segment quantizer. In the following, we note these briefly, with reference to Figs. 2.3 and 2.4.

It can be seen that scalar quantization (as in the basic LPC-10 vocoder [T82]) quantizes each co-efficient separately, while vector quantization quantizes the entire LP parameter vector. This is shown in row 2 in Figs. 2.3 and 2.4a. The principles of VQ that lead to rate-distortion advantage over scalar quantization are well established, such as based on exploiting the higher dimensionality of vectors, linear and non-linear dependencies between the vector components, and vector pdf shape [M85]. The VQ-LPC coder [W82, M85, WJC83] marked an important milestone in ultra low bit-rate coding by applying the then emerging concept of vector quantization (VQ) to quantize the LP parameters as a vector for each frame of speech as against the conventional scalar quantization of the parameters as in the standard LPC-10 vocoder [T82]. This brought about a remarkable reduction of the bit-rate from 2.4 Kbps to 800 bps while preserving the quality of the LPC-10 vocoder.

A natural extension of quantizing a vector to a vector-of-vectors (or also referred to as 'fixed length segments'), saw the emergence of matrix-quantization based LPC-10 system, which reduced the bit-rate to 300 bps while preserving the quality of coded speech as same as that of LPC-10 [WJC83, TG85]. This is shown as row 3 in Figs. 2.3 and 2.4b. Other variants of the basic matrix quantization concept have also been proposed, notably, the multi-frame coding schemes [KCT91, KCT92] at 600–800 bps, [ML92] at 800 bps, [MLG95] at 800 bps and the matrix product quantization [Br95] at 300 to 700 bps. Such multi-frame coding paradigms

2.1 Vector and Matrix Quantization

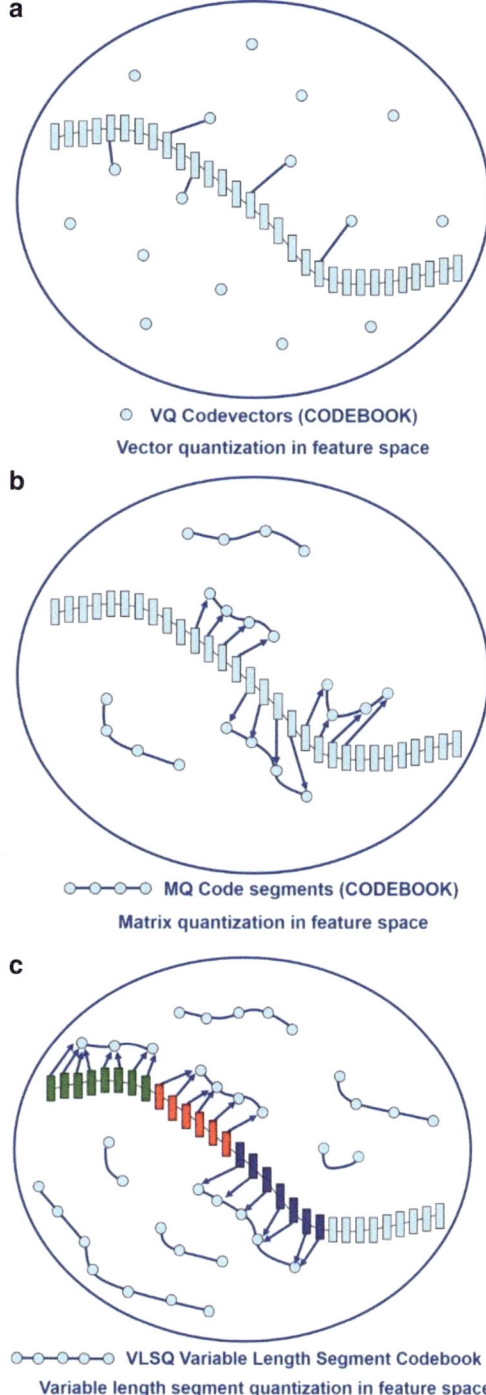

Fig. 2.4 Vector, matrix and variable length segment quantization in feature space

approximate a matrix of LP parameters (also called a super-frame) by a few frames as anchor points, with the other frames being differentially encoded or interpolated at the receiver. Among these, [Br95] employed a more rigorous formulation of representing the reconstruction matrix Y (of an input matrix X of m successive LPC parameter vectors to be quantized) as a product of a diagonal centroid matrix S representing the average parameter vector of the matrix and a temporal contour represented as a contour matrix V, and arriving at an iterative framework to design the joint S and V matrix codebooks. This work also extended this formulation to variable size matrices, i.e., segments, as is the topic of discussion in the next section. A further generalization of the multi-frame encoder of [KCT91] is the combined quantization-interpolation (CQI) method of [LF93] where the 'target' vectors defining the break-points or the end-frames of variable length segments marked by spectral discontinuity are determined sequentially in a two-pass framework to realize spectral distortions lower than that of [KCT91] at bit-rates below 350 bps. As we shall note later, the maximum-likelihood formulation of [SS87] in Sect. 2.2.1.2 provides an optimal framework for such a scheme, as was explored in and compared in [S94].

The next important development was the progression from quantizing 'fixed length segment' to 'variable length segments', as in [SH88]. These techniques exploited the variable durations of speech units (typically, phones) and then quantized them efficiently using structured or unstructured 'segment' codebooks. Segment vocoders based on variable-length segment quantization has provided the means of achieving low to ultra low bit-rates in the range of 800 to 150 bps while offering intelligible speech quality [RSM83, RWR87, SH88, HS92]. This is illustrated in row 4 in Figs. 2.3 and 2.4c. We shall consider this class of quantization in a more general setting that is referred to as 'segment vocoders', in the following section in a little more detail, and provide an overview of a wide variety of ultra low bit-rate quantizers in this framework.

2.2 Segment Vocoders

The basic functioning of a segment vocoder framework is given in Fig. 2.5 and can be described as made of the following main components and/or steps:

1. **Automatic segmentation**: Segmentation of input speech (a sequence of LP parameter vectors) into a sequence of variable length segments (also referred to as 'units').
2. **Variable-length segment quantization**: Segment quantization of each of these segments using a segment codebook $C = (c_1, c_2, \ldots, c_N)$ and transmission of the best-match code—segment index and input segment duration.
3. **Joint segmentation quantization**: While the early systems used automatic segmentation and segment quantization as separate steps, (i.e., segmentation yielded the variable length segments that are further quantized), most segment

2.2 Segment Vocoders

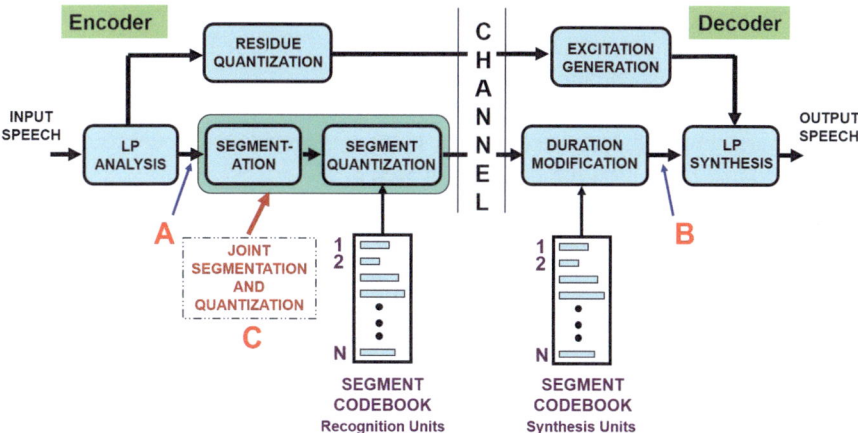

Fig. 2.5 Generic structure of a segment vocoder using LP analysis (at the encoder) and LP synthesis (at the decoder)

vocoders that evolved subsequently employed techniques that can be referred to as 'joint segmentation quantization', wherein the segmentation and quantization are combined within a single step for a given segment codebook, essentially solving what can be called a segmentation and labeling using methods derived from speech recognition. In Fig. 2.5, this is shown as block C that combines Step 1 and 2 above.

4. **Segment codebook:** The segment codebook of size N is typically made of N variable length segments or statistical models of segmental units such as HMMs, that are derived from training speech (long sequence of LP vectors) using special techniques such as joint segmentation and clustering procedures (or segmental K-means algorithm) and acoustic modeling techniques in speech recognition that yield the HMMs. Note that in Fig. 2.5, we have referred to the segment codebook at the encoder as 'recognition units' and the one at the decoder as the 'synthesis units'. As we will see in subsequent sections detailing this aspect, these are identical codebooks in most segment vocoders, particularly those using templates [RSM82b, RSM83, RWR87, RS04] and in the HMM based recognition/synthesis systems [T98, H03, MTK98]. However, these could also be different codebooks, differing in terms of the representation of the segments in the codebook (as long as each segment is associated with the same underlying acoustic properties); an example is [MBCC01, MGC01], where the recognition units are HMMs derived from LPCC representation, while the decoder has synthesis units that allow either LP synthesis or HNM synthesis. Other examples include [C08, CMC08a, CMC08b, PSVH10] (using syllable HMMs for segment quantization at the encoder and template based synthesis at the decoder) [LC01, LC02] (MFCC units for unit-selection based segment quantization at the encoder and HNM based synthesis at the decoder).

5. **Duration modification:** The code-segment corresponding to each received code-segment index is time normalized (duration modified) to match the duration of the corresponding input segment.
6. **Residual modeling and quantization:** The residual obtained by LP analysis is also parameterized, quantized and transmitted; the residual decoder reconstructs the residual to be used for synthesis in Step 6.
7. **Synthesis:** Synthesis of speech by LP synthesis using the code-segment time-normalized to match input segment duration.

The various segment vocoders proposed till date differ primarily in some or all of these aspects. In the following, we briefly describe each of the above components in more detail, primarily providing a review of how various segment quantization techniques for ultra low bit-rate coding differ with reference to these specific components and steps outlined above.

2.2.1 Automatic Segmentation

Here, we briefly review the automatic segmentation techniques in various segment vocoders that derive a variable length segmentation prior to segment quantization.

Much of the basic architecture in the segment vocoder framework was laid by the work at BBN [MCRK77, SKMKZ79, SKMS80, RSM82a, RSM82b, RSM83, RW85, RWR87, M06]. Some of these early work followed a segment-then-quantize approach, for example in [RSM82b, RSM83] where automatic segmentation was done using a simplistic spectral transition measure to generate diphone-like units which are used in both segment codebook design and segment quantization. In contrast, [RWR87] employed a combinatorial search for the best segmentation of a given block of input speech, under the constraint that segment durations are to be from a finite set, and optimized to minimize the quantization error for the block.

An interesting work by Svendsen [S94] showed that using an VQ codebook for quantizing segments derived automatically by the maximum-likelihood (ML) segmentation [SS87] yields a bit-rate reduction by a factor of 2, while preserving the speech quality. This clearly established the essence of segment quantization, namely, that it is more efficient to quantize a segment as whole, if the segment corresponds to an acoustic unit such as a phone (which comprises of a steady-state and therefore can be quantized parsimoniously by a single vector or a code-segment from a segment codebook).

[RS04] extended these early studies above and explored automatically derived phone and diphone units using spectral-transition measures and phone-like units using maximum-likelihood segmentation [SS87] and showed that automatically derived phone-like units are more efficient than diphone-like units for typical segment codebook sizes (made of randomly selected phone-like or diphone-like units in the form of segments).

2.2.1.1 Spectral Transition Measure

The 'spectral transition measure' (STM) is based on the principle of measuring the spectral derivative at every frame instant. STM was adopted in early segment vocoders for diphone-like segmentation [RSM82b, RSM83]. We consider two types of STM as used in [RSM82b], namely, the d_1 and d_3 measures. These are defined as follows: Let $\boldsymbol{o}_n$ be the LP parameter vector at the nth frame. The STM at frame n, $d_i(n)$, is given by $d_i(n) = \|\boldsymbol{o}_n - \boldsymbol{o}_{n-i}\|^2, i = 1, 3$.

Figure 2.6 illustrates the STM profile with respect to phone-like and diphone-like segments and the corresponding peak/valley picking as described below. $d_1(n)$ as a function of n exhibits peaks at fast spectral transitions (such as from one phone to another) and valleys at steady state regions (such as within a vocalic segment). $d_3(n)$ gives a smoother measure of the spectral derivative. Thus, peak-picking of $d_1(n)$ or $d_3(n)$ locates transitions or phone boundaries and results in a phone-like segmentation. Picking the minima (valleys) of these functions locates a frame within steady-state regions that has maximum local stationarity and corresponds to a diphone boundary. Successive peaks therefore mark phone-like (PL) segments and successive valleys mark diphone-like (DPL) segments.

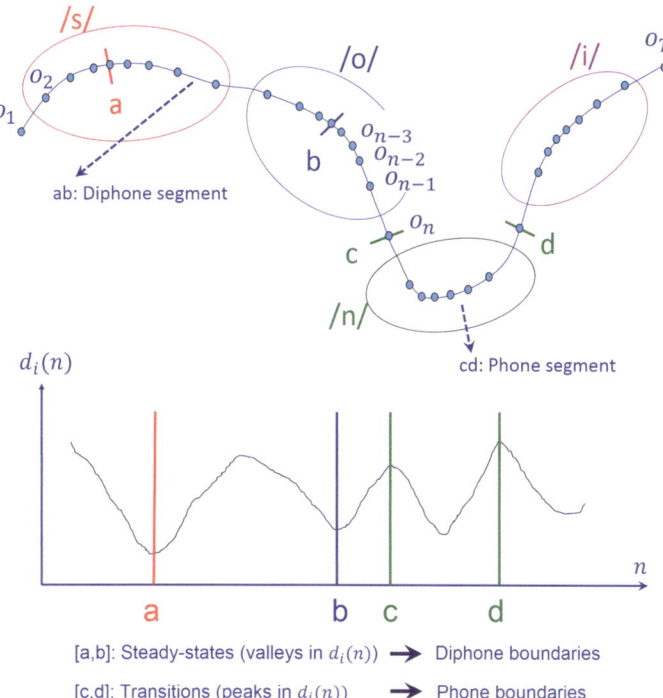

Fig. 2.6 Spectral transition measure (STM) based definition of phone-like and diphone-like segmentation and corresponding STM profile used for peak/valley picking

[RS04] used an extrema picking algorithm (EPA) for peak- and valley-picking on $d_1(n)$ and $d_3(n)$ functions. This algorithm employs a threshold (δ) to detect the extrema (peaks and valleys) alternatingly in a left-to-right scanning. The algorithm can be stated as follows:

<u>Search</u>
for a peak (valley) by repeated updating of current maximum p (current minimum v) every time a local maximum (minimum) is detected
<u>Until</u>
a function value smaller than $(1-\delta)p$ (larger than $(1+\delta)v$) is encountered.
<u>After this,</u>
Go to <u>Search</u> *and start searching for a valley (peak).*

While small values of δ (close to 0) result in over-segmentation, large values of δ (close to 1) result in under-segmentation and needs to be optimized for a desired segment rate. Thus, the threshold δ used in the extrema-picking-algorithm plays a crucial role in determining the quality of segmentation and hence needs to be optimized to yield good segmentation match as defined above. For this purpose, [RS04] used the segment rate (number of segment/second) as the primary measure to be matched. For instance, TIMIT has a phone-rate of $R=12.5$ phones/s, as measured over 300 sentences. In STM, [RS04] set δ to a value that results in this segment rate. The optimal δ corresponding to a segment rate of 12.5 segments/s also results (automatically and interestingly) in the highest percent match (between the automatically determined segment boundaries and the true segment boundaries within a specific tolerance limit) as well as the lowest percent insertion and deletion values. Based on segmentation applied on TIMIT sentences, it was shown that d_1 is well suited for detecting fast spectral transitions such as phone boundaries and hence in better phone-like segmentation than d_3. Conversely, the smoother d_3 is more suited for a diphone-like segmentation than d_1, where it is necessary to detect steady-states by valley-picking. As a consequence, [RS04] used STM (d_1) for phone-like (PL) segmentation and STM (d_3) for diphone-like (DPL) segmentation in the overall vocoder.

2.2.1.2 Maximum-Likelihood Segmentation

Let a speech utterance be given by $O_1^T = (o_1, o_2, \ldots, o_T)$, which is a LP parameter vector sequence of T speech frames, where, o_n is a p-dimensional parameter vector at frame 'n'. The segmentation problem is to find 'K' consecutive segments in the observation sequence O_1^T. Let the segment boundaries be denoted by the sequence of integers $B = (b_0, b_1, \ldots, b_{k-1}, b_k, \ldots, b_K)$. The kth segment starts at frame $b_{k-1}+1$ and ends at frame b_k; $b_0=0$ and $b_K=T$. This is illustrated in Fig. 2.7.

The maximum likelihood (ML) segmentation is based on using the piecewise (quasi) stationarity of speech as the acoustic criterion for determining segments. The criteria is to obtain segments which exhibit maximum acoustic homogeneity within

2.2 Segment Vocoders

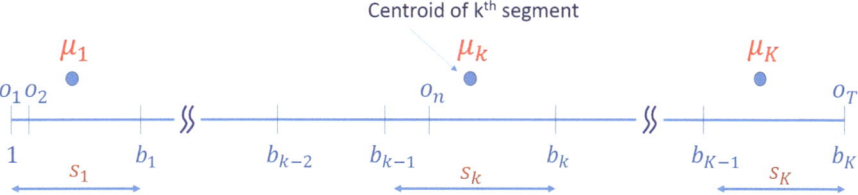

Fig. 2.7 Schematic of maximum-likelihood segmentation

their boundaries. The acoustic inhomogeneity of a segment is measured in terms of an 'intra-segmental distortion', given by the sum of distances from the frames that span the segment, to the centroid of the frames comprising the segment. For a given K, the optimal segmentation $B^* = (b_0^*, b_1^*, \ldots, b_K^*)$ is obtained so as to minimize the sum of intra-segment distortion over all possible segment boundaries, i.e., minimize

$$D(K,T) = \sum_{k=1}^{K} \sum_{n=b_{k-1}+1}^{b_k} d(o_n, \mu_k)$$

which can be given as,

$$B^* = \arg\min_B D(K,T)$$

where, $D(K,T)$ is the total distortion of a K-segment segmentation of $O_1^T = (o_1, o_2, \ldots, o_T)$; μ_k is the centroid of the kth segment consisting of the spectral sequence $O_{b_{k-1}+1}^{b_k} = \{o_{b_{k-1}+1}, \ldots, o_{b_k}\}$ for a specific distance measure $d(\cdot,\cdot)$. For the Euclidean distance 'd', μ_k is the average of the frames in the segment $O_{b_{k-1}+1}^{b_k}$. This is illustrated in Fig. 2.7.

The optimal segment boundaries are solved efficiently using a dynamic programming (DP) procedure [SS87, SRS02] using the recursion

$$D(k, b_k) = \min_{b_{k-1}} [D(k-1, b_{k-1}) + \Delta(b_{k-1}+1, b_k)]$$

for all possible b_{k-1}; $D(k, b_k)$ is the minimum accumulated distortion up to the kth segment (which ends in frame b_k), i.e., $D(k, b_k)$ is the minimum distortion of a segmentation of $(o_1, o_2, \ldots, o_{b_k})$ into k segments; $\Delta(b_{k-1}+1, b_k)$ is the intra-segment distortion of the kth segment $O_{b_{k-1}+1}^{b_k}$. This is illustrated in Fig. 2.8. The segmentation problem is solved by invoking (2) for $D(K,T)$; this is computed efficiently by a trellis realization. The optimal segment boundaries $(b_0^*, b_1^*, \ldots, b_K^*)$ are retrieved by backtracking on the trellis along the optimal alignment path corresponding to $\min\{D(K,T)\}$.

Figure 2.9 shows a schematic of the feature space where the input feature sequence corresponding to the word 'emotion' is segmented by ML-segmentation into corresponding phone-like units.

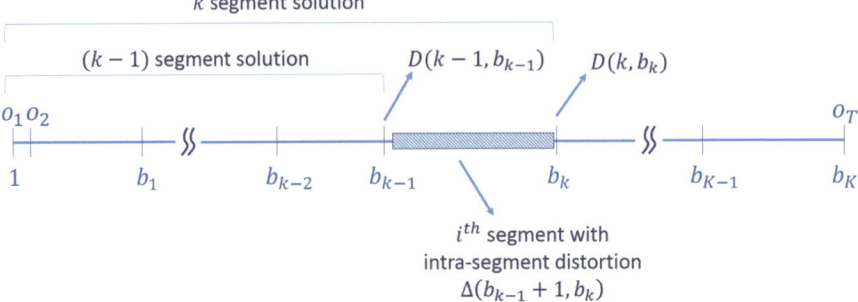

Fig. 2.8 Recursive structure employed in maximum-likelihood segmentation

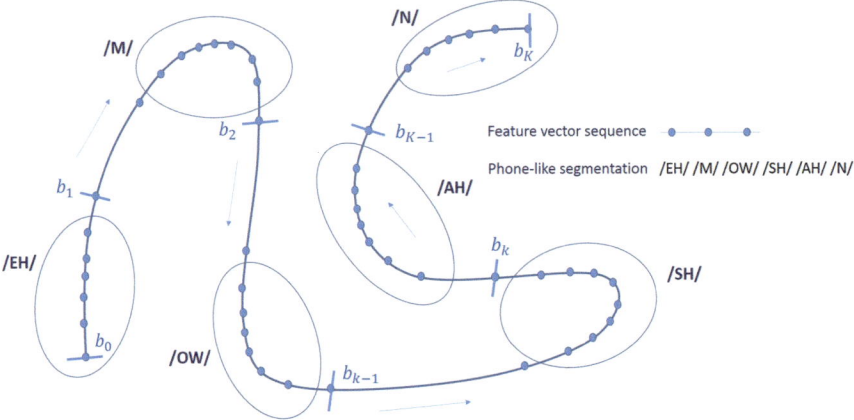

Fig. 2.9 Illustration of a typical phone-like segmentation derived by ML segmentation in a feature space

2.2.1.3 ML Segmentation: Duration Constrained (ML(DC))

By definition, ML segmentation produces a segmentation where each segment is maximally homogenous; when the segment rate equals the phone-rate of natural speech, the resulting segments will be quasi-stationary and would correspond to the steady-state regions of phonetic units. However, even for a correct segment rate, ML segmentation can result in segment lengths which are unnaturally short (1 frame long) or long (up to even 70 frames). Such segments will be distorted significantly during segment quantization and result in poor vocoder performance.

In the distribution of phone durations in TIMIT database, nearly 95 % of the labeled phonetic segments are in the range of 1-20 frames. In order to limit the segment lengths of ML segmentation to such a meaningful range (of actual phones durations), [RS04] modified the ML segmentation to have 'duration constraints'. Here, the optimal segments are forced to be within a duration range of $[\alpha, \beta]$, where

2.2 Segment Vocoders

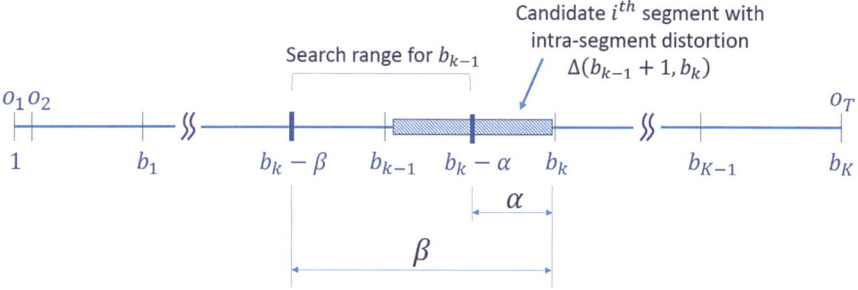

Fig. 2.10 Recursive structure employed for duration-constrained ML segmentation

Fig. 2.11 Different segmentation techniques in STM and ML framework and corresponding phone-like and diphone-like units explored in [RS04]

	Segmentations	
	STM	ML
Units	d_1 d_3	UC DC
PLU	○ ○	○ ○
DPLU	○ ○		

UC: Unconstrained
DC: Duration-constrained

α and β are the minimum and maximum lengths possible (in frames). Segments of lengths $<\alpha$ and $>\beta$ are not generated at all. This is achieved by restricting the candidate boundaries in the search for optimal segment boundaries in (2) as follows, and as illustrated in Fig. 2.10:

$$D(k, b_k) = \min_{b_k - \beta \leq b_{k-1} \leq b_k - \alpha} \left[D(k-1, b_{k-1}) + \Delta(b_{k-1} + 1, b_k) \right]$$

This also has the advantage of reducing the computational complexity of ML segmentation from $O(T^2)$ to $O(lT)$, where $l = \beta - \alpha + 1$ with typical values of $[\alpha, \beta] = [2, 20]$.

This work [RS04] evaluated the STM and ML segmentations for phone-like units (PLU) and diphone-like units (DPLU) as shown in Fig. 2.11. It used various segmentation accuracy measures and segmental SNR to compare the performance of the different phone-like and diphone-like units (by retaining the residual without modeling or quantization), and showed that ML segmentations realize phone-like units which are significantly better than those obtained by STM in terms of match accuracy with TIMIT phone segmentation as well as actual vocoder performance measured in terms of segmental SNR. Further, it was shown that the phone-like units of ML segmentations also outperform the diphone-like units obtained using STM in early vocoders. The resulting segment vocoder had an average bit-rate of 300 bps, with very high intelligibility when used in a single-speaker mode. The interesting observation to be made in this context is that phone-like units are typically characterized by segments between two transitions with a stable steady-state at its center. Hence these units have a good means of representing a phonetic unit for a given codebook size, despite the high invariability arising in the transitional parts arising from context dependency. In contrast, the diphone units, defined from the steady-state of one phoneme to the steady-state of another, has low variability and hence represent a particular diphone better than a phone-like unit would. However, considering that the number of diphones is large $O(M^2)$ for M phones, the codebook size needs to be larger than that of phone-like units to ensure adequate coverage and representation of all possible diphones. Thus, for a given codebook size of the order of 8,192, as considered in this work, the phone-like units have the possibility of being represented by several segments belonging to a given phone-class (e.g. 128 segments = 8,192/64, for number of phones as 64) thereby accounting for contextual variability to a certain extent; on the other hand, this codebook size is still inadequate for representing diphone-like units (e.g. $64^2 = 4,096$), which approximately allows only two segments of a diphone class to be present in the codebook. (Of course, these numbers are approximate to allow reasoning with the scenarios, and the units are bound to be distributed non-uniformly). This gives a reasonable explanation for why the above work reported a better performance of the phone-like units for the kind of codebook sizes employed.

2.2.1.4 ML Segmentation: A Generalized Basis

An interesting observation to note here is that the ML segmentation formulation outlined above sets the basis for at least five different kinds of segmentation and quantization techniques that are discussed in this chapter, namely,

1. Distortion constrained segmentation and vector quantization of [S94] pointed to in Sect. 2.2.1 (also generalizing over the sequentially determined combined quantization-interpolation method of [LF93] indicated in Sect. 2.1) and discussed in some detail under the R/D optimal techniques in Sect. 2.3.4.
2. Joint segmentation and quantization technique forming the core of the variable length segment quantization [SH88], and derived based on the 2-level dynamic programming algorithm discussed in Sects. 2.2.3, 2.2.3.2 and 2.2.3.3.
3. R/D optimal algorithm of [PV00] discussed in Sect. 2.3.1.
4. Rate-distortion constrained segmentation in the variable-to-variable length vector quantization (VVVQ) formulation of [CL94] discussed in Sect. 2.3.2.
5. Multigram formulation (and sequence segmentation) of [B95] and reinterpretation with reference to VVVQ in [BCC97], discussed in Sect. 2.3.3.

2.2.1.5 Syllable-like units and other segmentations

In an interesting departure from classical units such as phone-like and diphone-like units, [C08, CMC08a, CMC08b, PSVH10] proposed the use of syllable as the unit of segmentation and quantization and derive syllable-like units using group-delay based automatic segmentation [NM04, MY11] in a syllable-based segment vocoder framework. Other work to use syllable based unit definition for segmentation are [CKBC99] which detects local maxima of the function of 'sonority decrease' to further identify and select syllables, (this work uses syllables as alternative to the ALISP units, discussed separately in Sect. 2.5), and [HN89] which uses syllable as an unit for recognition and synthesis, and defines reference patterns of syllable units using hand labeling.

In a study of segmentation techniques for segment vocoders, [C05] considers the segmentation algorithm by [v91, BS00], reporting spectral distortions of 2.5–1.5 dB at bit-rates in the range of 100–400 bps. [v91] define a segmentation by treating phonemes as stationary regions and detecting the transitional boundaries of a phoneme marked by a drop in the normalized correlation between frames below a specified threshold. [BS00] use a vowel-spotting approach by peak-detection of average magnitude, by viewing vowels as characterized by larger average magnitudes than nearby frames.

2.2.1.6 Temporal Decomposition

Temporal decomposition is an important technique [A83] to model the articulatory dynamics of speech production, where speech is described as a result of a sequence of distinct articulatory gestures towards and away from articulatory targets, the influence of the neighboring gestures thereby contributing to the co-articulatory effects commonly underlying all continuous speech. Atal [A83] proposed temporal decomposition as a means of describing a sequence of parametric vectors representing continuous speech as a linear combination of the underlying targets, by solving for a set of target vectors which when linearly combined by (a small number of) overlapping target functions, can approximate a parametric vector at each frame. Among the further work in this direction [V89, VM89, BA91], Van Dijk-Kappers [V89] studied the phonetic relevance of temporal decomposition. Temporal decomposition has evinced interest for application in speech recognition and speech coding, notably very low bit-rate speech coding, given its appealing property to approximate continuous speech in terms of a parsimonious set of representations in the form of the target vectors and target (overlapping) functions, even while being amenable to synthesis by the very nature of the formulation being to approximate each vector in the sequence by the linear combination of a limited number of target vectors. In this regard, we note here that the temporal decomposition framework has led to significant work in the context of very low bit-rate speech coding, such as the 450–600 bps vocoders of [CS90, CS91, CS93] yielding speech, that, when compared to the 2,400 bps LPC, is very intelligible and more natural sounding speech, the 600 bps range vocoder yielding intelligible speech in [GD96, GDB97b], comparative study of different spectral representations of [GDB97a] including LPC, reflection coefficients, LAR, cepstrum, and band filters for event detection with respect to their decomposition suitability (not for their phonetic relevance, but the degree of reconstruction accuracy possible), 1,000–1,200 bps high fidelity compression in [GDS98], the 996 bps vocoder of [Sung98] using temporal decomposition on LSF parameters yielding reasonable speech quality compared to the 2,400 bps LPC10e and a related work of [P95] in terms of step decomposition of speech. An important application of temporal decomposition is in the definition and modeling of ALISP units, which has been successfully applied for very low bit-rate speech coding, and which has been discussed separately in Sect. 2.5.

In the context of deriving segments by an automatic segmentation algorithm, it is to be noted that the problem of automatic segmentation to yield various types of meaningful acoustic segments is well studied with numerous techniques proposed till date. While the above sections did touch on some of the salient segmentation techniques specifically in the context of ultra low bit-rate speech coding, a broader discussion into all other techniques is beyond the scope here. Details of a wide range of automatic segmentation techniques can be found in the very comprehensive review of Vidal and Marzal [VM90].

2.2.2 Segment Quantization

Segment quantization as an explicit step is applicable only when the automatic segmentation is done as a first step to result in variable-length segments, and hence is specific to only those techniques discussed above under automatic segmentation.

Once the variable length segments are obtained, each of these segments s_k, $k = 1, \ldots, K$ is quantized using the best segment c_{q_k} in the segment codebook $C = (c_1, c_2, \ldots, c_N)$ defined as the segment yielding the lowest segmental distortion. This is shown in Fig. 2.12.

Quantizing an input segment to the best matching segment from the segment codebook calls for the definition of a distance metric to measure the segmental distortion between two arbitrary variable-length segments (the input segment to be quantized and a segment from the segment codebook). Such a measure needs to account for non-linear temporal and acoustic variability between two such segments, and ideally, this is best computed by dynamic time warping (DTW) well defined and applied for speech recognition and widely used in the ALISP units based segment quantization discussed in detail in Sect. 2.5. Linear warping, as a computationally simpler approximation to a full-fledged DTW as well as the so-called space-sampling based matrix-distortions have also been applied.

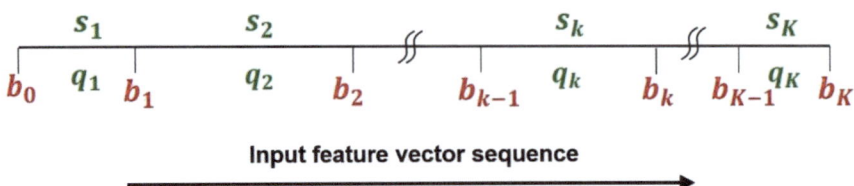

Fig. 2.12 Segmentation and quantization

2.2 Segment Vocoders

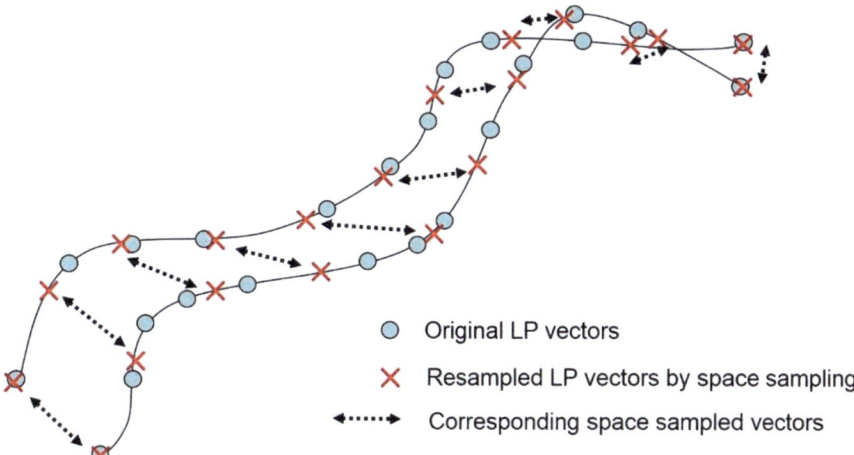

Fig. 2.13 Schematic of space sampling and correspondence between space-sampled vectors between two trajectories

For instance, in the primary work from BBN [RSM82b, RSM83, RWR87] which performed a diphone-like segmentation, they employed a space-sampling technique by which both the input segment and each of the variable-length segment in the segment codebook were re-sampled in the feature space (e.g. LAR vector space) to yield a fixed number of vectors that are equally spaced along the trajectory of the original segments. This can be visualized as shown in Fig. 2.13. The number of resampled vectors is set as 10, thereby converting each variable length segment to a fixed length segment (matrix) of length 10. Subsequently, a matrix distortion is computed as the sum of 10 Euclidean distortions, each between the corresponding LAR vectors of the resampled input segment and codebook segment. An input segment is then quantized to the codebook segment that has the minimum matrix distortion. While being computationally simple, this method circumvents a more exact and optimal dynamic time-warping kind of distance calculation, whose computation cost is proportional to the lengths of the input segment and each segment in the segment codebook being matched. The latter work of [RS04] also followed the same approach for segment quantization using a segment codebook for phone-like and diphone-like units.

In the case of the segment codebook being a HMM codebook, i.e., each entry is an HMM modeling a phone or phone-like or acoustic units, segment quantization takes the form of finding the best matching HMM from the HMM codebook, that has the maximum likelihood (usually, the Viterbi likelihood) for the input segment. While most of the HMM codebook based methods perform a transcription of the input speech in terms of the units that the HMMs model (and thereby falling under the joint segmentation and quantization approach, described in the next sub-section), the syllable based approach of [C08, CMC08a, CMC08b, PSVH10] used syllable HMM codebook, and performed a separate segment quantization using this method of finding the best matching HMM for an input segment, which is a syllable-like segment derived from a first stage group-delay based segmentation, as pointed out in Sect. 2.2.1.

2.2.3 Joint Segmentation Quantization

We first give the problem definition of joint segmentation and quantization, before attempting a brief overview of different techniques that falls within this definition. In reviewing the class of segment vocoder that use joint segmentation and quantization as above, we also point to the nature of the segment codebook employed in the various joint segmentation and quantization work alongside here, considering the dependence of the joint segmentation and quantization (or segmentation and labeling) on the 'segment codebook'. We also devote a separate section for segment codebooks in order to highlight additional details as may be required in specific instances.

2.2.3.1 Basic framework

As the name implies, joint segmentation and quantization performs a segmentation and quantization of the associated segments simultaneously, i.e., given an input sequence of speech feature vectors (e.g. LPC parameter vectors), it obtains a set of segments each labeled by a segment from a segment codebook under some optimality consideration. This can be viewed as a 'segmentation and labeling' process, where each segment is labeled by a segment index from a segment codebook. Due to this, joint segmentation and quantization necessarily requires a segment codebook with which the segmentation and labeling is done.

Figure 2.12 shows a typical segmentation and labeling of a feature vector sequence $O = (o_1, o_2, \ldots, o_T)$. Let the segment codebook used for such a segmentation and labeling be given by a collection of N variable length segments $C = (c_1, c_2, \ldots, c_n \ldots, c_N)$, where a segment c_n is a sequence of LP parameters of some length. Figure 2.12 shows the segmentation of the input feature vector sequence into K segments, defined by the segment boundaries $B = (b_0, b_1, \ldots, b_K)$. Each segment $s_k = (x_{b_{k-1}+1}, x_{b_k})$ is given a label q_k, where $q_k \in (1, N)$. Let the set of labels associated with the K segments $S = (s_1, s_2, \ldots, s_{k-1}, s_k, \ldots, s_K)$ be $Q = (q_1, q_2, \ldots, q_{k-1}, q_k, \ldots, q_K)$.

A typical segmentation is completely defined by

1. The number of segments K,
2. The segment boundaries for a given K, $B = (b_0, b_1, \ldots, b_{k-1}, b_k, \ldots, b_K)$ and,
3. The segment labels $Q = (q_1, q_2, \ldots, q_{k-1}, q_k, \ldots, q_K)$.

2.2 Segment Vocoders

The problem of joint segmentation and quantization naturally needs to minimize the quantization error (or distortion) associated with each segment, and hence a measure of total segment distortion which, for a given (K, B, Q), can be expressed as

$$D = \sum_{k=1}^{K} d(s_k, c_{q_k})$$

The optimal segmentation is determined as (K^*, B^*, Q^*) that minimizes D. Thus the solution for optimal segmentation can be given as

$$(K^*, B^*, Q^*) = arg \min_{K,B,Q} D = arg \min_{K,B,Q} \sum_{k=1}^{K} d(s_k, c_{q_k})$$

This is clearly a combinatorial search problem, which by brute force search will require first hypothesizing a K, and then hypothesizing a B with K segment boundaries, and for each resulting segment, obtaining the best matching codebook segment c_{q_k} and repeating this for all K and B, until the minimum overall distortion D is found. Fortunately, this problem is already well solved in the context of speech recognition, where the above definition of segmentation and labeling is exactly what corresponds to the now well established 'connected word recognition' (CWR) problem [RJ93], where we are given a input feature vector sequence of a test speech spoken in a continuous fashion made of a sequence of words drawn from a fixed vocabulary of N words and it is required to determine the sequence of words spoken, i.e., that best explains the feature vector sequence according to a distance measure or likelihood measure with respect to a set of N word models that the CWR system has [which could be template models or hidden Markov model (HMM)]. The CWR problem has at least three types of algorithmic solutions [RJ93], namely,

(a) 2-level dynamic programming (DP) algorithm
(b) Level-building algorithm
(c) One-pass DP algorithm

In light of the above basic definition of joint segmentation and quantization, it can be noted that 'segment quantization' techniques in segment vocoders essentially performed a 'connected segment recognition' by which they determined the optimal segment boundaries (and hence the segment lengths) and the segment labels that were transmitted and used for reconstructing speech at the decoder after duration normalization.

2.2.3.2 Shiraki and Honda Variable-Length Segment Quantization

At this point, it is appropriate to refer to the work of Shiraki and Honda [SH88, HS92] as who first established the notion of joint segmentation and quantization in the context of variable length segment quantization. While the joint segmentation and quantization algorithm used by Shiraki and Honda [SH88] was essentially the 2-level DP algorithm proposed much earlier by Sakoe [S79] for connected word recognition, the importance of the variable length segment quantization work in [SH88] lies in the fact that [SH88] defined and addressed the design of 'optimal' variable length segment codebooks of any desired size, from continuous speech training data using an iterative 'joint segmentation and clustering' algorithm, which in speech recognition literature had been developed independently and referred to as the segmental K-means (or SKM) algorithm [RWJ86, RJ93]. We shall illustrate this codebook design algorithm in Sect. 2.2.4 as part of a discussion on 'segment codebook'. [SH88] also defined segment to segment distortions (within the joint segmentation and quantization framework) using linear interpolation and warping transform matrices.

2.2.3.3 2-Level DP Framework for Joint Segmentation and Quantization

We give now a brief note on the 2-level DP algorithm employed by Shiraki and Honda [SH88] for performing joint segmentation and quantization, particularly with respect to how it can be viewed and realized as a variant of the ML algorithm described in Sect. 2.2.1.2. It was seen that the ML algorithm solves for the optimal segmentation boundaries as B^* given by,

$$B^* = \arg\min_{B} D(K, T)$$

where $D(K, T)$ is the sum of the K intra-segmental distortions each being defined as in the second summand below,

$$D(K, T) = \sum_{k=1}^{K} \sum_{n=b_{k-1}+1}^{b_k} d(o_n, \mu_k)$$

In the ML formulation, the intra-segment distortion for a segment $O_{b_{k-1}+1}^{b_k} = \{o_{b_{k-1}+1}, \ldots, o_{b_k}\}$ is defined with respect to the centroid μ_k of this segment. Instead, assume that the ML formulation is to be posed as the joint segmentation and quantization problem, as defined above, such that each resulting segment has the lowest possible quantization distortion and the segments (from the solution for the segment boundaries) are to be determined so as to minimize the total distortion. In this case, we first need to take into account a segment codebook from which the best segment is to be chosen to optimally quantize a given segment. The ML formulation can now be redefined so as to yield the desired joint segmentation and quantization with respect to this explicitly defined codebook by performing two changes in the basic ML equations:

(a) The intra-segment distortion $\sum_{n=b_{k-1}+1}^{b_k} d(o_n, \mu_k)$ is replaced by the minimum quantization distortion of the segment $O_{b_{k-1}+1}^{b_k} = \{o_{b_{k-1}+1}, \ldots, o_{b_k}\}$ with respect to the segment codebook $C = (c_1, c_2, \ldots, c_n \ldots, c_N)$, i.e., by defining $D(K,T)$ as

$$D(K,T) = \sum_{k=1}^{K} d_k^*$$

where, d_k^* is the minimum quantization distortion of the segment $O_{b_{k-1}+1}^{b_k} = \{o_{b_{k-1}+1}, \ldots, o_{b_k}\}$ given by

$$d_k^* = \min_{c_n \in C} d\left(O_{b_{k-1}+1}^{b_k}, c_n\right)$$

where $d\left(O_{b_{k-1}+1}^{b_k}, c_n\right)$ is the segmental distortion between segment $O_{b_{k-1}+1}^{b_k}$ and a code segment c_n in the segment codebook C, realized by an appropriate time-warping (such as an optimal dynamic time-warping [RJ93]) or by the linear warping defined by [SH88].

(b) Note that in the ML formulation, the overall distortion $D(K,T)$ decreases monotonically with K, reaching a value of 0 when $K=T$, i.e., the input feature vector sequence of T vectors is segmented into T segments, yielding the trivial solution of each segment being a single vector. However, interestingly, in the case of 2-level DP based segmentation and quantization (or connected segment recognition), when $D(K,T)$ is obtained for a range of K of interest (from considerations of minimum to maximum search range for the desired number of segments, in the extreme case giving the limits of 1 to T), it is possible to obtain an optimal K^* for which $D(K^*,T)$ is a minimum, i.e., the segment to segment matching requirement dictates and constrains the segmentation solution to an optimal number of segments for which the overall quantization distortion is minimum, with less or more number of segments leading to duration mismatches between the segments in the segment codebook and the input segments being quantized. This leads to the second modification to the ML formulation, that K^* is obtained as

$$K^* = \arg\min_{K} D(K,T)$$

Thus the 2-level DP can be realized as a reformulation of the ML-segmentation algorithm with the above two modifications, defined with respect to an explicit codebook (in the place of the implicitly defined 'centroid' as the approximation of a segment) with the joint segmentation and quantization solved as (K^*, B^*, Q^*) as above (Note that $Q^* = q_1^*, q_2^*, \ldots q_k^*, \ldots, q_{K^*}^*$ is obtained incidentally from K^* and

B^* as solved above, where once segment s_k is solved as $s_k = \left\{ O_{b_{k-1}^*+1}, \ldots, O_{b_k^*} \right\}$, the quantization index q_k^* for this segment is simply given as

$$q_k^* = arg \min_{n=1,\ldots,N} d\left(O_{b_{k-1}^*+1}^{b_k^*}, c_n \right)$$

2.2.3.4 One-Pass DP Algorithm

We now briefly describe the one-pass DP algorithm listed as one of the three algorithms for solving the connected word recognition (or 'connected segment recognition' in the context of joint segmentation and quantization) for two reasons—firstly that it is also called the now well-known Viterbi-decoding when the word model is a HMM and we shall shortly discuss phone HMM based phoneme recognition which forms the joint segmentation and labeling for a class of segment quantizers, and secondly, we will use the one-pass DP algorithm in a modified form in the subsequent chapters to realize the unified and optimal algorithms for unit-selection based segment quantization and this short introduction to this algorithm serves as a precursor to these following chapters.

Here, we will mainly illustrate how the solution looks like for the one-pass DP algorithm for the case when the segment codebook is a set of templates of variable length segments, rather than go into the actual algorithmic realization of the one-pass DP algorithm (which can be found in [RJ93, N84]). The one-pass DP algorithm is applied on the sequence of feature vectors (in the x-axis) and the set of templates in the segment codebook (in the y-axis), and derives the solution in the form of an optimal decoding path, as shown in Fig. 2.14, which is retrieved by a backtracking procedure at the end of applying a path growing procedure (made of recursions) from left to right (i.e., from the starting frame in the input feature vector sequence to the last frame). This path establishes the mapping between the input feature vector sequence and the templates, yielding the segmentation and labeling solution (K^*, B^*, Q^*) as follows: The discontinuities in the path represent a transition from one word to another, and the number of distinct such sub-paths yields the optimal number of segments K^*. The segment to segment transition, marked by the discontinuity, corresponds to the optimal segment boundaries B^* and the identity of the sub-path (mapping a segment in the input feature vector sequence), in terms of the template q_k in the y-axis, gives the optimal label to the corresponding input segment s_k. The one-pass DP solution also yields the optimal matching distortion D^* corresponding to this optimal path, which is a measure of the overall quantization distortion in the joint segmentation and quantization.

2.2.3.5 Phoneme Recognition and Phonetic Vocoders

A typical application of the above 'connected word recognition' is the continuous phone recognition problem, where the input speech (in the form of a feature vector

2.2 Segment Vocoders

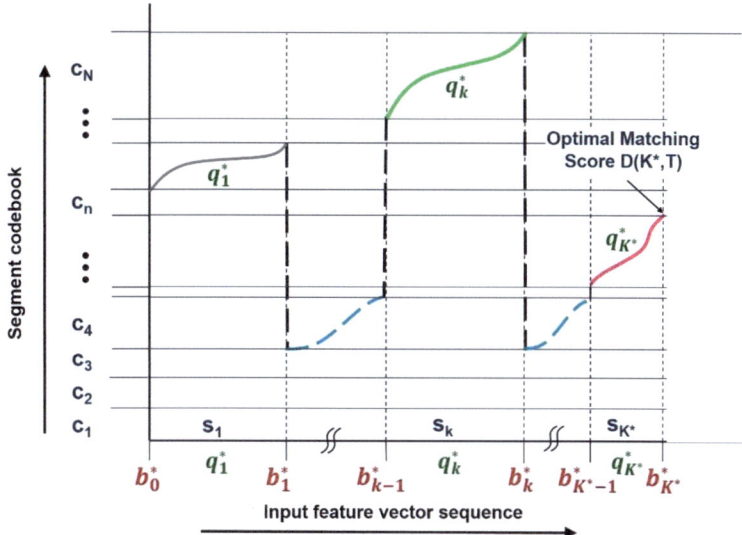

Fig. 2.14 Joint segmentation and quantization (segmentation and labeling) in an one-pass DP (Viterbi) framework

sequence) is decoded into a sequence of phonemes using a set of phone-models, typically HMMs, either in the form of monophone (context independent) models or triphone (context-dependent) models. This will be seen to be one of the primary joint segmentation and quantization technique repeatedly used in a good number of segment vocoders to derive phone-like segmentation and transmission of phonetic indices.

Though we attributed the advent of joint segmentation and quantization framework to Shiraki and Honda's variable length segment quantization algorithm and the joint segmentation and clustering algorithm for design of variable length segment codebooks, the principle of performing a 'connected phoneme recognition' using phone HMMs dates back to the work of [MCRK77, SKMKZ79, SKMS80] from BBN. These work represent the earliest work in segment vocoder framework focused on deriving and transmitting phonetic units at the encoder, using a phoneme recognition system and fall within the framework of joint segmentation and quantization. In this work, the transmitted phone indices were then used in the decoder to derive a diphone sequence that were used for synthesis, following an earlier feasibility study that established the merit of diphone synthesis that alleviates concatenation issues that is typical of phone-based concatenation and synthesis.

The next work (and an early one representing a conceptual milestone in ultra low bit-rate coding), that also employed phone recognition in the encoder (as in [SKMS80] referred above) and thereby falls within the joint segmentation-quantization framework is the 'phonetic vocoder' system of [PD89]. Here, the input speech is segmented and labeled in terms of phonetic units using speech recognition techniques (essentially a connected phone recognition system) with the phone HMM acoustic models (numbering 60) trained from phonetically labeled TIMIT database. The state sequence path, as a sequence of state transition information, through each phone HMM in the resulting phone transcription is further used to specify the frame-level quantization of the input feature vector (10th order LPC coefficients), which are transmitted and used at the decoder to reconstruct the speech signal.

To generalize further, an entire class of segment vocoders evolved based on the principle of phone recognition using HMM based acoustic modeling of phonetic or acoustic sub-word units, and using conventional speech recognition techniques (for deriving a phonetic transcription of the input speech to be coded) [IP97, OPT00, T98, H03, MTK98, McC06]. All of these undoubtedly fall within the joint segmentation and quantization framework, by virtue of performing a phoneme recognition, implying a simultaneous solution to both the segmentation and labeling problem as defined above. Having noted this class of segment vocoders, we shall discuss the HMM based approaches separately in some detail in a section to follow (Sect. 2.4).

2.2.4 Segment Codebook

As a precursor to discussing variable length segment codebooks, it is best to consider the fixed length variants, namely, the vector quantization (VQ) codebooks and matrix quantization (MQ) codebooks. As the name implies, VQ is a codebook of single vectors, derived by the K-means algorithm or the LBG algorithm [M85]. MQ is derived similarly, by treating each matrix segment as an entity and defining matrix-to-matrix distortion, and setting the problem of MQ codebook design also in the K-means framework [M85]. The interesting work of Svendsen [S94], who showed that a ML segmentation can be made to yield a variable length segmentation, with each segment quantized by a single vector from a VQ codebook also fits into the notion of a vector codebook, but applied on variable length segments. In a similar vein, the R/D optimal linear prediction, which uses a collection of LP models (of various orders and quantized and coded in different ways) to quantize variable length segments, also fits into the notion of a vector codebook, though derived in a manner similar to variable-length segment quantizer design, but with additional constraints on the rate of the resulting codebook, which we shall show later (in a separate section devoted to R/D optimal LP coding, Sect. 2.3).

Segment codebooks that have been employed in segment vocoders typically fall in two categories:

(a) Templates: Variable length segments, also called templates, made of a sequence of feature vectors (e.g. LP parameter vectors, MFCCs, etc.) representing phones, diphones, syllables or automatically derived acoustic units.

(b) Hidden Markov models (HMMs): Parametric models of phones (context-independent monophones or context-dependent triphones), diphones, syllables or automatically derived acoustic units.

These type of segment codebooks have been used both for an independent segment quantization (following a first stage automatic segmentation) as well as in a joint segmentation and quantization step (as is more common for HMM based modeling of phonetic units in the form of phoneme recognition).

2.2.4.1 Template Segment Codebooks

The segment codebooks, in the form of templates are either derived as randomly populated segments (selected from a long sequence of speech feature vectors by means of automatic segmentation, such as described in Sect. 2.2.1 [RSM82b, RSM83, RWR87, RS04]) or designed by means of the joint segmentation and clustering algorithm (or the segmental K-means algorithm) as in [SH88], as pointed out earlier in Sect. 2.2.1. Considering the importance of the VLSQ scheme using the clustered codebook design, we add some details on this here. Figure 2.15 shows a typical joint quantization and clustering algorithm (also referred to as segmental K-means algorithm), which shares it iterative framework with the more conventional K-means algorithm, but now adapted to segments occurring in continuous speech, with the added step of having to extract them optimally in the first place, followed by defining a cluster of such segments and its centroid. This algorithm is illustrated in Fig. 2.15 and explained in the following.

The joint segmentation and clustering algorithm (or the SKM algorithm) comprises the following two steps carried out iteratively, until convergence determined typically by a rate of decrease in the overall quantization distortion over the training corpus:

1. **Joint segmentation and quantization**: This step first performs a segmentation and labeling of the input speech (training corpus) using an initial segment codebook, which could be randomly populated fixed length segments or variable length segments derived from some other automatic segmentation methods. This segmentation and labeling is carried out by the joint-segmentation and quantization algorithm (realized in the form of a 2-level DP algorithm, as noted in Sect. 2.2.3.3) as shown in Fig. 2.15a in the form of the input feature vector sequence being segmented and labeled in terms of the segment codebook indices, for e.g. the variable length segments (A, B, C, D, E, F, G) labeled with the index 32 are shown. For a segment codebook of size N, this step results in N clusters of variable length segments each indexed by one of the codebook segments. Each such cluster has a number of variable length segments which can be said to share the same acoustic property as the codebook segment whose index it is labeled by.
2. **Segment codebook update**: In this step, each codebook segment (corresponding to a cluster derived above) is replaced by the centroid (or also called the pseudo-

Fig. 2.15 Joint segmentation and clustering algorithm (or the segmental K-means algorithm). (**a**) Basic iterative structure, (**b**) Centroid update

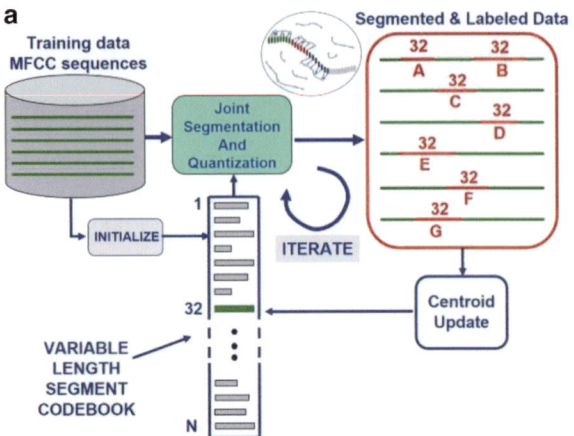

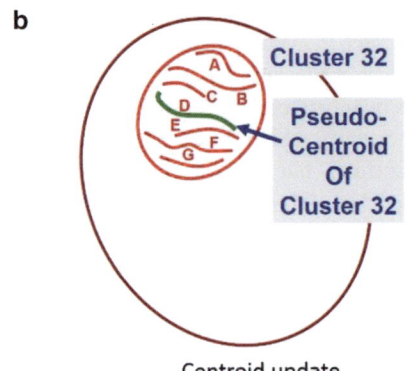

centroid) of the cluster, defined as the segment (from among the segments in the cluster) which has the lowest average segment-to-segment distortion with all the other segments in the cluster. When defined as a pseudo-centroid, the resulting optimal segment satisfying this condition happens to be one of the segments in the cluster. This is illustrated in Fig. 2.15b, which shows a cluster corresponding to index 32, made of the variable length segments (A, B, C, D, E, F, G). The centroid update step chooses segment D as the pseudo-centroid (that minimizes the average intra-cluster quantization distortion) as the one that replaces the previous codebook segment 32. Once all the N codebook segments are updated in this manner, the new codebook is used to perform a variable length segmentation in Step 1.

The above two steps are carried out iteratively, until the average segment quantization distortion (for instance, as is obtained during the centroid update step, as the average of the N intra-cluster segment quantization distortion with respect to the updated codebook) converges, i.e., the rate of decrease of this segment quantization distortion (on the training corpus) converges below a threshold. The segment codebook at the end of the convergence is the desired variable length segment codebook.

2.2.4.2 HMM Segment Codebook

The segment codebooks, in the form of HMMs (say, phone HMMs) are derived by now well established techniques in speech recognition, namely, acoustic model training. Here, the HMMs are trained from a training corpus (which could be manually or semi-automatically) labeled in terms of the units of interest (phones, diphones, syllables, etc.), e.g. as in [T98, H03, MTK98, C08, CMC08a, CMC08b, PSVH10]. Alternately, and more commonly, the training speech corpus would have only the associated orthographic transcription from which model-estimation algorithms such as 'embedded re-estimation' are used to train the HMMs. As will be indicated in a following section on HMM based recognition-synthesis techniques, an interesting departure from phone HMMs are the HMM modeling of abstract acoustic units (automatically derived units) corresponding to multi-grams which can span several successive phonetic units [CBC98a] or the recognition acoustic units (RAUs) and synthesis acoustic units (SAU) of [BC03].

2.2.5 Duration Modification

As noted in Fig. 2.5, the decoder receives the indices of segments (from the segment codebook) which quantize successive segments in the input (derived by either separate automatic segmentation and segment quantization or a single-step joint segmentation-quantization). These indices are used to retrieve the corresponding segments from a segment codebook, which is usually identical to the one in the encoder or identical in the nature of the underlying speech each segment represents or models (in the case of HMM), but represented using different parametric representation more conducive for synthesis. The retrieved segments are concatenated together and used for synthesis along with the residual information (also received by the decoder in some parameterized and quantized form) in the case of a LPC synthesis framework or other prosodic information in the case of alternate synthesis framework (such as HNM, PSOLA or HTS).

In the case of template based segment codebook representation, an important step in the decoder 'prior' to the concatenation of the templates corresponding to the received indices is 'duration modification' (also called 'duration normalization'), where a template in the segment codebook corresponding to a received index is modified to have a duration of the original segment in the input speech (at the encoder). The durational information is also quantized and transmitted to the decoder, thereby enabling this duration modification step. This has been typically

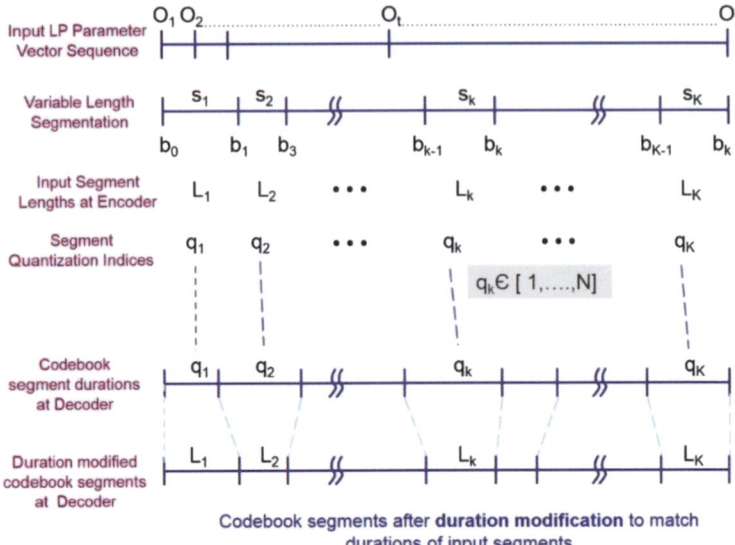

Fig. 2.16 Illustration of duration modification in the flow of processing starting from segmentation and quantization (at encoder) and duration modification codebook segments (at decoder)

done via space-sampling [RSM82b, RSM83, RWR87] (as depicted in Sect. 2.2.2) or by a linear warping matrix [SH88] or DTW warping path [CBC00]. In the case of the segment codebook being a HMM codebook, the duration modification is implicitly carried out during HMM-based synthesis using the state-duration information transmitted as side-information [PD89, T98, H03, MTK98].

The general notion of duration modification is shown in Fig. 2.16. The input LP parameter vector sequence $o_1, o_2, \ldots, o_T$ is segmented into K segments $(s_1, s_2, \ldots, s_{k-1}, s_k, \ldots, s_K)$ of duration $(L_1, L_2, \ldots, L_{k-1}, L_k, \ldots, L_K)$ and corresponding segment quantization labels (or indices) $(q_1, q_2, \ldots, q_{k-1}, q_k, \ldots, q_K)$. The encoder transmits both the segment indices $(q_1, q_2, \ldots, q_{k-1}, q_k, \ldots, q_K)$ and the corresponding original segment durations $(L_1, L_2, \ldots, L_{k-1}, L_k, \ldots, L_K)$. At the decoder, let the segment codebook of size N (identical to the one at the encoder) be $C = (c_1, c_2, \ldots, c_n \ldots, c_N)$ with corresponding lengths $(l_1, l_2, \ldots, l_n \ldots, l_N)$. Each of the segments in the segment codebook which is retrieved using the received indices thus has its own duration, i.e., a segment in the codebook c_{q_k} retrieved using index q_k is supposed to represent the input segment s_k; the length of input segment s_k is L_k, whereas the length of the corresponding codebook segment c_{q_k} is l_k. This necessitates the duration modification of the codebook segment c_{q_k} (of duration l_k) to duration L_k so as to match the duration of the input segment s_k. Let this duration modified codebook segment be c'_{q_k}; the sequence of duration modified codebook segments at the decoder is now $\left(c'_{q_1}, c'_{q_2}, \ldots, c'_{q_{k-1}}, c'_{q_k}, \ldots, c'_{q_K}\right)$ and speech synthesized from these concatenated units will have the individual units matching the corresponding units in input speech and together match the duration of the original speech.

2.2.6 Residual Parameterization and Quantization

Figure 2.5 showed the general structure of a segment vocoder in the framework of LP analysis and synthesis. In this framework, the spectral quantization part enjoys most attention given the importance of minimizing spectral distortion for high quality synthesis at the decoder. However, the role of the residual information in this LP framework is equally important, as is evidenced by the fact that the original LPC-10 vocoder is acknowledged to have a synthetic voice owing to the use of a voiced/unvoiced modeling of the residual in terms of a pulse train or random noise as the excitation signal as an approximation of the original residual in synthesis at the decoder. Much effort has been expended in rendering the LPC-10 quality better by an enhanced modeling of the residual, such as in the MELP coder [M08].

In the context of the segment vocoder framework shown in Fig. 2.5, the residual therefore continues to play an important role in the overall quality of synthesized speech at the decoder. A wide variety of residual modeling, parameterization and quantization has been adapted in the segment vocoder literature, each motivated by both parsimonious representation of the residual and ensuring minimal loss of speech quality during synthesis.

The basic modeling and representation of the residual in terms of voicing/unvoiced decision and pitch (with the gain parameter considered here as a global parameter, governing the short-term energy of the signal) continues to be the primary means of quantization in early segment vocoders, though in each case employing some special means of lowering the effective rate for the residual (e.g. 800 bps VQ-LPC uses 8 bits/frame for pitch, gain and voicing quantization by combining three consecutive frames in comparison to 13 bits/frame in the 2,400 bps LPC-10 vocoder). Specific techniques for incurring highly reduced bit rates for pitch or gain information typically includes modeling and quantizing pitch and gain profiles over an entire segment (the segments determined by automatic segmentation or joint segmentation quantization step) or use of vector quantization to quantize such pitch and gain vectors. For instance, we note here some salient techniques used in the residual (for LP based synthesis) or prosody (for other types of synthesis, e.g. HNM) modeling and quantization. In the earliest segment vocoder work, [RSM82b, RSM83] transmitted pitch with only 3 bits/segment and 1 bit/segment respectively, using a piece-wise linear approximation of pitch, with the pitch profile in a segment being modeled by a linear function, and the change in pitch quantized by an adaptive 2-level quantizer, with the level being proportional to the segment duration. The latter work on variable length segment quantization [SH88] used 4 bits/segment for differential pitch quantization.

Considering the more recent HMM based recognition-synthesis system, [T98] used pulse train or white noise excitation (of the MLSA filters) as in the classic LP vocoder, but reported results without quantization of pitch. In a subsequent work in the same framework, [H03] used a $F0$ quantization scheme based on a VQ version of the multi-space distribution HMM. Piece-wise linear approximation of the contour was adapted in other work too, such as in [LC99], in a rate-distortion framework for coding the pitch contours and notably in their later unit-selection framework [LC01, LC02], though used in a HNM based synthesis at the decoder. From among the series of work reported based on the ALISP units based on HMM recognition and synthesis by LPC or HNM, [PCB04a, PCB04b] can be seen to use unit selection based on a pitch profile correlation between the input segment and codebook segments, and quantizing a pitch correction parameter to apply on the codebook segment to match with the input segments pitch profile. In the parallel formant synthesizer used in the decoder in [OPT00], an appropriate excitation signal is chosen, depending on the amount of voicing and frication present in the sound being synthesized. [McC06] explored an interesting combination of the HMM based phonetic recognition and MELP based residual modeling within a predictive vector quantization of the MELP parameters for synthesis to realize a scalable phonetic vocoder framework operating at bit rates from 300–1100 bps.

2.2.7 Synthesis

As indicated in Fig. 2.5, the segment vocoder is typically set in LPC vocoder framework based on the source-filter model, with LP analysis in the encoder computing the spectral parameters (filter part) and the residual (source part) and quantizing and transmitting them. The synthesizer at the decoder uses the LPC synthesis framework to use the received filter and residual parameters to synthesize speech. When the residual is modeled by voicing decision, pitch and gain, the synthesized speech resembles that of LPC-10 vocoder. This is more or less the synthesizer framework and methodology adapted in a range of segment vocoders [W82, RSM82a, RSM82b, RSM83, RWR87].

A major departure from this was adapted in the unit-selection based coders of Lee and Cox [LC01, LC02], who employed the 'harmonic plus noise model' (HNM) framework, though continuing to use prosodic parameter estimation at the encoder, and their use in modifying the synthesis unit's representation for HNM based synthesis at the decoder. The ALISP unit based work (reported in a series of papers which are reviewed in Sect. 2.5) [MBCC01, MGC01], explored both LPC synthesis framework and the HNM framework, choosing to prefer HNM over LPC synthesis, owing to the artifacts observed in LPC synthesis.

In an interesting departure from these work, [OPT00] used a parallel formant synthesizer at the decoder, driven by synthesis parameters derived from rules which in turn were acquired by a mapping from acoustic segments obtained by the speech recognition based segmentation at the encoder. In what can be considered as a convergence of recognition-synthesis frameworks, the HMM recognition-synthesis work of [T98, H03, MTK98] used the HMM-based speech synthesis framework, now widely popular, successful and referred to as HTS [T13], where the synthesis is done using the MLSA filter (derived from the mel-cepstral coefficients derived from the concatenated state sequence of the recognized HMM phone units) and driven by a pulse train or white noise depending on the voicing decision. The fact that the quality of the HTS synthesis in a speaker-adaptive mode could reach a MOS of 4 [MTK98] seems to herald a new direction in ultra low bit-rate coding, where high quality speech is possible in a speaker-adaptive (and hence speaker-independent) mode of operation.

2.3 R/D Optimal Linear Prediction

In this section, we review a class of segment quantization algorithms, namely by Prandoni and Vetterli [PGV97, PV00], Chou and Lookabaugh [CL94] and the closely related work of Baudoin et al. [BCC97], all set within a formulation of rate-distortion optimization, where the segmentation and quantization is performed with respect to constraints on the overall rate of the resulting quantization, as well as impacting the underlying codebook design. We also provide an interpretation of Svendsen [S94], which sets a constraint on the distortion in performing the segmentation.

2.3.1 Prandoni and Vetterli R/D Optimal Linear Prediction

An important development in the framework of segment quantization is that of Prandoni et al. [PGV97] and Prandoni and Vetterli [PV00] termed R/D optimal linear prediction. To state briefly, in this work, segmental distortion is defined with respect to the LPC residual error and the bit-rate is defined as the trading-off parameter based on the cost of the LPC order for each segment, so as to yield a R/D trade-off in controlling the choice of LP order per segment and the resulting overall distortion, resulting in ultra low rate coding (working in a variable bit-rate manner) with average bit-rates of 300–900 bps.

Considering the importance of such a formulation in the segment vocoder framework, we give below in some detail the basic formulation of this work, but using the notations already introduced above in the definition of the variable length segment quantization framework in Sect. 2.2.3. A codebook is given in the form of a collection of N different LP models (which could be predictors of different orders as well as predictors whose parameters are quantized and coded in different ways). Let this codebook be $C = (c_1, c_2, \ldots, c_n \ldots, c_N)$. Let a segmentation and quantization of an input speech LP parameter sequence $O = (o_1, o_2, \ldots, o_T)$ be given by the segment boundaries $B = (b_0, b_1, \ldots, b_{k-1}, b_k, \ldots, b_K)$, with corresponding segments

$(s_1, s_2, \ldots, s_{k-1}, s_k, \ldots, s_K)$ and the quantization labels $Q = (q_1, q_2, \ldots, q_{k-1}, q_k, \ldots, q_K)$. Here, the quantization label q_k represents a LP model c_{q_k} (of some order) from the codebook C. The distortion corresponding to such a segmentation and quantization is given by $D(O, K, B, Q)$ given by the sum of the K segmental LP prediction errors $d^2(s_k; c_{q_k})$, where each of these errors is the squared LP prediction error when model c_{q_k} is applied on segment s_k, i.e.,

$$D(O, K, B, Q) = \sum_{k=1}^{K} d^2(s_k; c_{q_k})$$

In addition, this work defines the overall bit-rate for such a quantization by using a cost function $r(c_{q_k})$ that reflects the cost (in bits) as a function of the order $b(c_{q_k})$, including the side information for specifying the segment duration and the relative LP order; i.e., the overall bit-rate load for a typical segmentation of O in terms of (K, B, Q) is given by

$$R(O, K, B, Q) = \sum_{k=1}^{K} r(c_{q_k})$$

The important formulation in this work arises from solving the segmentation and quantization problem to yield the optimal (K^*, B^*, Q^*) under a rate-distortion tradeoff specified by the constrained minimization for a given K, given by

$$(B^*, Q^*) = \min_{B} \min_{Q} D(O, K, B, Q) \tag{2.1}$$

$$R(O, K, B^*, Q^*) \leq R^* \tag{2.2}$$

where Eq. (2.2) dictates that the overall quantization distortion $D(O, K, B, Q)$ be minimized under the constraint that the overall bit-rate load be less than a specified bit-rate budget (or limit) R^*. This formulation is easily understood by noting that the segmentation and quantization solution of Eq. (2.1) for a given K is simply the ML kind of segmentation with the intra-segment distortion of a segment being replaced by the squared LP error with respect to the 'best' LP model from the collection of possible LP models C. As with the ML formulation, this overall distortion is monotonically decreasing for increase in K, as the segments becomes shorter and shorter. However, when it is required that the overall bit-rate be limited to a maximum of R^*, this translates into the maximum number of segments that can be derived and further to the maximum order of the LP model that can be used for quantizing a segment; i.e., without the bit-rate budget constraint, an extreme solution is one where the number of segments equals the number of frames ($K = T$), i.e., each frame is modeled by its best LP model from the collection, which in turn can be the LP model of the highest order available. However, with the bit-rate budget in place, the number of segments is optimized with the bit-rate

2.3 R/D Optimal Linear Prediction

resource distributed over the segments, so that each segment is modeled by some LP model of lower order (not necessarily the highest order among the models available). This can also be seen as a combination of two functions as a function of the number of segments K, one being the overall distortion which decreases with K and the overall bit-rate which increases linearly with K.

When the constrained minimization problem in Eqs. (2.1) and (2.2) is combined via a Langrangian multiplier, these two functions combine to yield an optimum (minimum) for some K^*. Such a joint functional using a Lagrange multiplier is given by,

$$J(\lambda) = D(O, K, B, Q) + \lambda R(O, K, B, Q)$$

which can further be simplified to yield the optimization as

$$J^*(\lambda) = \min_B \min_Q \left\{ \sum_{k=1}^{K} d^2(s_k; c_{q_k}) + \lambda r(c_{q_k}) \right\}$$

It can be easily seen that this is exactly in the form of a ML segmentation that can be reformulated and solved by the 2-level DP algorithm outlined above in Sect. 2.2.3.3, by defining the distortion (or cost) associated with the kth segment [in the recursion of Eq. (2.3)] as the term

$$\min_{c_{q_k} \in C} \left\{ d^2(s_k; c_{q_k}) + \lambda r(c_{q_k}) \right\}$$

for some choice of λ (which weighs the bit-rate load in addition to the distortion function appropriately). The recursion, as in the 2-level DP algorithm, is given by

$$J^*_{[1,b_k]} = \min_{1 \leq b_{k-1} \leq b_k - 1} \left\{ J^*_{[1,b_{k-1}]} + \min_{c_{q_k} \in C} \left\{ d^2(s_k; c_{q_k}) + \lambda r(c_{q_k}) \right\} \right\} \quad (2.3)$$

where $J^*_{[1,b_k]}$ is a k segment segmentation of the input segment $O_1^{b_k} = \{o_1, \ldots, o_{b_k}\}$ described recursively as the sum of the optimal $k-1$ segmentation of the segment $O_1^{b_{k-1}} = \{o_1, \ldots, o_{b_{k-1}}\}$ (i.e., $J^*_{[1,b_{k-1}]}$) and the minimum cost (segmental squared LP error of segment $s_k = O_{b_{k-1}+1}^{b_k} = \{o_{b_{k-1}+1}, \ldots, o_{b_k}\}$ plus the corresponding λ weighted bit-rate load) associated with segment s_k. The final solution is obtained by invoking the above recursion [Eq. (2.3)] for $J^*_{[1,N]}$ for a given λ and finding the optimum over a desired range of K. Note that the solution is obtained in the same manner as the 2-pass DP kind of recursion solved by a trellis realization, and recovering the optimal segmentation and quantization parameters by backtracking.

2.3.2 Variable-to-Variable Length Vector Quantization

In the context of designing segment codebooks within the constraint of a rate-distortion trade-off, a related work is that of Chou and Lookabaugh [CL94]. This is an important work in the following ways: It formulated the problem of 'variable-to-variable' length codes, where the terminology derives from whether the segments (or blocks) in the input speech (parameter sequence) vary in length or not and whether the blocks of channel symbols vary in length or not. The term 'variable-to-variable' refers to a scheme that quantizes variable length segments into variable length bit patterns (channel symbol). While it is interesting to note that this formulation handles variable length segments in the input speech, the 'variable' part of the channel symbol comes from performing an entropy coding (e.g. Huffmann code) that exploits the non-uniform probability distribution of the segment usage from a segment codebook. While this formulation is set in a constrained rate-distortion optimization (as in [PV00] above), the solution derived is very similar to the one by [PV00] in the form of a recursive relation that can be seen as a variant of the ML-formulation (Sect. 2.2.1.2) redefined into the 2-level DP solution. [CL94] also give an iterative solution for the design of the segment codebook, along the same lines as the joint segmentation and clustering (but with the additional bit-rate constraint on the joint segmentation and quantization step) or the segmental K-means algorithm as noted above in Sect. 2.2.4.1.

An important aspect in this formulation is that, though the segment codebook is made of variable length segments, the mapping between an input segment to the codebook segment that quantizes the input segment is such that they are of same length, i.e., it does not perform any kind of segment to segment warping between potentially unrestricted set of variable length codebook segments to each input segment as may be required. [CL94] points to this as an important difference from Shiraki and Honda's [SH88] variable length segment quantization algorithm. As a result of not using segment to segment warping, [CL94] show that their method outperforms that of [SH88]. Moreover, [CL94] note that a related work Jeanrenaud and Peterson [JP91, PJV90] from BBN, is more close to their formulation, in the sense that [JP91] use duration dependent segment codebooks, which is populated (and designed by means of an iterative algorithm) with segments of variable lengths, but ensuring that it has sub-codebooks each with several segments of a fixed length. However, it should be noted that having a variable length segment codebook and allowing for warping based segment quantization allows a particular segment representing a particular acoustic realization of speech to quantize a varied number of input segments belonging to the same acoustic category; in the absence of such a warping of a particular segment in the codebook, the codebook is constrained to have segments of different lengths to match with the different realizations of the same acoustic category (i.e. without warping), thereby increasing the effective codebook size and the bit-rate to achieve a desired quantization distortion.

2.3.3 Multigrams Quantization

In a related work, Baudoin et al. [BCC97] provided a reinterpretation of the VVVQ method of Chou and Lookabaugh [CL94] with reference to the multigrams quantization method, where a spectral vector sequence is segmented and quantized using a codebook of variable length segments called multigrams, and the codebook is obtained by maximizing the joint likelihood of the optimal segmentation and the observation sequence under two different scenarios, one when in the form of quantization indices (classical multigrams) and the other without vector quantization (modified multigrams) [BCC97]. The concept of multigrams was earlier introduced in an interesting formulation by Bimbot et al. [B95], which allowed segmentation of a symbol string (e.g. text data or spectral sequence indices resulting from VQ) into characteristic sequences, and building a dictionary of such sequences using a maximum likelihood parsing of a string W using Viterbi decoding, given by

$$L(W) = \max_{\{B\}} \prod_k p(S_k)$$

where, B is the set of all possible segmentations into variable length sequences S_k. The solution to this parsing is expressed in a recursive manner akin to the maximum likelihood (or the 2-level DP algorithm) formulation and solved by dynamic programming.

2.3.4 Distortion Constrained Segmentation

In the context of the above R/D optimal segmentation, it can be noted that the earlier work of Svendsen [S94] can be restated in the form of R/D constrained segmentation, but with the difference that in this work, the segmentation was optimized to let the distortion be not above a threshold distortion (corresponding to the 1 dB transparent quality quantization), and the associated segmentation was observed to reduce the bit-rate by a factor of 2, i.e.., the number of segments could be reduced to the extent that each segment can be made of several frames thereby ensuring more effective use of a codevector, and such that when each segment is quantized by the best codevector in a vector quantization codebook, the average distortion is limited to be less than the specified threshold. This is illustrated in Fig. 2.17. Note that point A corresponds to the baseline case of each frame being quantized by a codevector in the VQ codebook; the corresponding distortion is not 0, as was the case with ML segmentation (for the limiting case when the number of segments K equals the number of frames T in the input sequence and where the centroid happened to be the approximating vector), since here the quantization of the individual frames is done with an external vector codebook

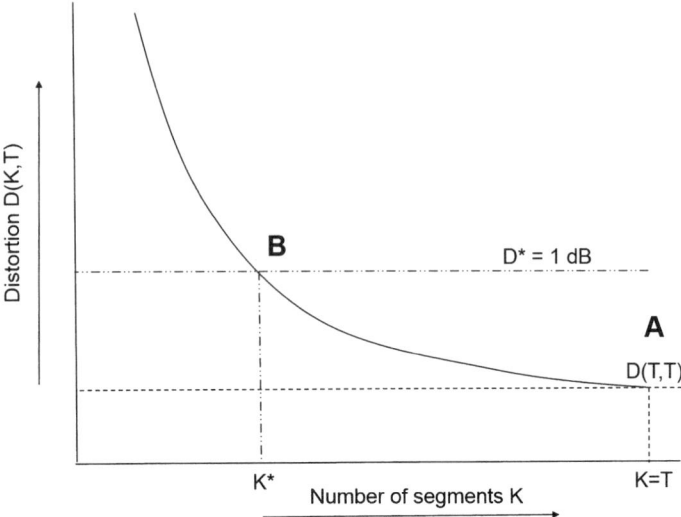

Fig. 2.17 Distortion vs. number of segments in the distortion constrained segmentation (in ML segmentation formulation) of [S94]

thereby yielding a minimum non-zero distortion, which sets the asymptotic performance, when $K = T$. Point B corresponds to the solution obtained by the algorithm of Svendsen [S94] characterized by an average distortion not greater than the specified threshold and the corresponding bit-rate derived from the associated number of segments (effective bit-rate = segment-rate $K^*/\left(\frac{T}{F}\right)$ segments/second × $\log_2 N$ bits/segment) for a VQ codebook of size N and a frame-rate of F frames/second.

The ML formulation of [SS87], adopted in [S94] as above in a distortion-constrained segmentation can also be seen to generalize over the combined quantization-interpolation (CQI) [LF93], referred to in Sect. 2.1, by noting that the intra-segmental distortion of a segment (as defined in the sub-section on ML-segmentation in Sect. 2.2.1.2) needs to be defined as the distortion due to approximating the frames in the segment by interpolated and quantized frames using the end frames of the segment as the 'target' vectors as defined in [LF93], even while retaining the constraint of realizing a distortion within some predefined threshold.

2.4 HMM Based Recognition-Synthesis Paradigm

As noted already in Sects. 2.2.3 and 2.2.4, joint segment quantization using phone-HMM codebooks for realizing a phonetic transcription of input speech represents a major departure in segment vocoder framework from the more classical segment

2.4 HMM Based Recognition-Synthesis Paradigm

codebooks in which each segment is a template comprising a sequence of LP parameter vectors and an associated template based segment quantization scheme. Following the early work of [SKMS80, PD89], the use of HMM codebooks and phonetic transcription has further lead to a class of segment vocoder systems that have refined the HMM based approach into a recognition/synthesis framework and have resulted in very high quality speech coding. We briefly review these here.

2.4.1 HTS Based Framework

Subsequent to the above early work that explored phonetic HMMs in a phonetic vocoder setting, a major initiative in HMM based segment vocoders emerged with the very successful paradigm of parametric speech synthesis, or HMM-based synthesis, now referred to as HTS [T13]. We review these briefly here [T98, H03, MTK98] which were originally referred to as being set in a 'recognition-synthesis' paradigm. Figure 2.18 shows a generic structure of the HMM based recognition-synthesis framework adapted in these work.

Tokuda et al. [T98] proposed an HMM-based speech recognition and synthesis technique for a very low bit rate speech coder operating at 150 bps and yielding subjective quality comparable to that of a VQ at 400 bps (50 frames/s and 8 bits/frame) and a DMOS of 3.4. This system derives the name of being in 'recognition-synthesis' paradigm, as the encoder is a HMM based phoneme recognizer and the decoder does the inverse operation of using an HMM-based speech synthesis

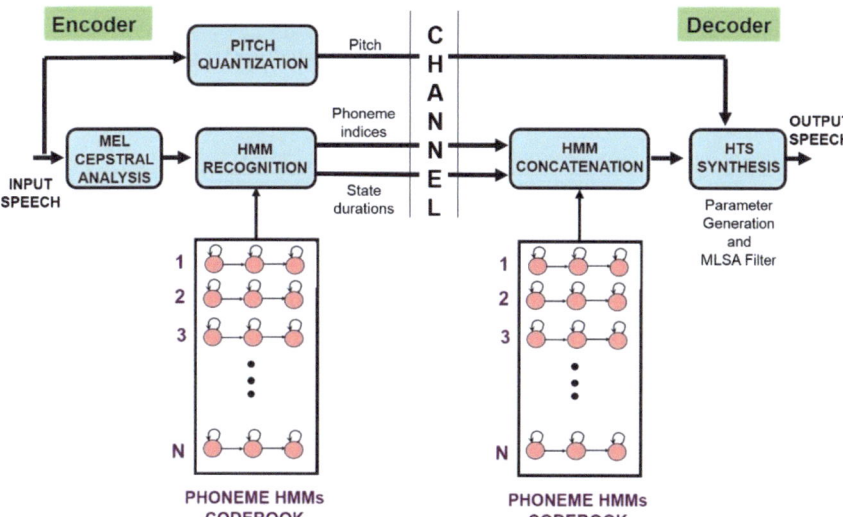

Fig. 2.18 Generic structure of HMM based recognition/synthesis framework

technique. More specifically, here, the speech spectra is represented by mel-cepstral coefficient vectors and transcribed into a phoneme sequence using context dependent (triphone) phoneme HMMs, where each triphone HMM was a 3-state left-to-right model with no skip. A total of 34 phonemes and silent models were used. Subsequent to phoneme recognition at the encoder using these triphone HMMs, the phoneme indices are transmitted to the decoder along with the state duration and pitch information. In the decoder, the phoneme HMMs corresponding to the phoneme indices are extracted from the phoneme-HMM codebook and concatenated and a sequence of mel-cepstral coefficient vectors (conditioned on the state sequence reconstructed from the received state duration information) is generated using a 'speech parameter generation' algorithm first proposed for HMM based speech synthesis [T13]. An important consequence of the parameter generation result is that the mel-cepstral coefficient vectors thus obtained reflect not only the means of static and dynamic feature vectors but also their covariances, resulting in a natural-sounding synthetic speech. Subsequent to the derivation of the mel-cepstral coefficients from the concatenated HMM, speech is synthesized by exciting a MLSA (Mel Log Spectrum Approximation) filter (derived from the mel-cepstral coefficients) by a pulse train or white noise generated according to the pitch information. However, a drawback of this coder was that it was speaker dependent.

Further improvements of the above phoneme-based HMM vocoder were proposed by Hoshiya et.al. [H03]. It was found that, by using a pitch coding scheme for quantizing each of the phoneme segments of the low bit-rate speech coder, the coder's performance at 110 bit/s was superior to that of a 600 bit/s VQ based vocoder in terms of the subjective MOS quality.

In a variant of these 'phonetic vocoder' approaches, [CCTC97] proposed a recognition-synthesis coder using HMM based phone modeling (48 phones and 2463 diphones) for phoneme recognition of the input speech to transmit the phonetic indices and further synthesis at the decoder using time-domain pitch-synchronous overlap method (TD-PSOLA) to realize a 750 bps coder which preserved speaker characteristics and had a MOS of 3.0.

2.4.2 Speaker Adaptive HMM Recognition-Synthesis

In order to render the first HMM-based recognition-synthesis vocoder speaker-independent, Masuko et al. extended the work in [T98] to be speaker adaptive in [MTK98]. The decoder was initially populated only with synthesis units from a single speaker. Therefore, regardless of the input speaker, the synthetic speech was limited only to the HMMs used in the decoder. Furthermore, it was likely that the mismatch between the training and input speakers caused recognition errors, further degrading the quality of the output speech quality. This problem could be handled in two ways. One was to have speaker dependent codebooks. The other more

2.4 HMM Based Recognition-Synthesis Paradigm

reasonable approach was to adapt a standard codebook to the input speech by accounting for any mismatch between the input speech and the HMMs trained at the decoder. As a result, the HMMs at the decoder are adapted to the input speech by moving the output distributions of HMMs by adding a transfer vector to mean vectors to fit the distributions to the input parameter vectors. The transfer vector is determined for each phonetic segment by maximizing the output probability of the input sequence.

In this work, the segment codebook comprised 49 speaker independent, 5-state left-to-right triphone HMM. Transfer vectors are also trained for every 100 ms of speech contributing an additional bit-rate of 100 bps. The speaker dependent (SD) coder of [MTK98] was shown to significantly improve the quality of the coded speech and the recognizability of the speaker as opposed to the speaker independent (SI) system in [T98]. For one of the speakers in the system, a Degradation Mean Opinion Score (DMOS) of 3.8 was reported for the SD coder and 4.0 for the speaker adaptive system, while the DMOS for an SI coder was 1.5. This remarkable enhancement in DMOS for speaker adaptation can be considered as a milestone in ultra low bit-rate speech coding, given that the overall coder now becomes applicable to unseen test speaker, even while being capable of yielding high DMOS, in keeping with the underlying HMM-based synthesis system's potential to yield high quality synthetic speech, which currently marks the success of the HTS framework for TTS [T13].

2.4.3 Ergodic HMM Framework

In a related, but conceptually different, work within the HMM-based recognition-synthesis paradigm, [Lee05] proposed use of an ergodic HMM to replace the set of phone-HMMs, by having a large number of states (e.g. 64) to model any spoken language as a sequence of abstract acoustic units, which correspond to the individual states of such an ergodic HMM. By definition, such an ergodic HMM is trained from a long unlabeled training corpus and enjoys the advantage of not requiring phonetically labeled training corpus as is required for the phone-HMMs of [T98, H03, MTK98]. A transcription of input speech using such a large ergodic HMM yielded a state sequence which is transmitted to the decoder along with fractional pitch. The synthesis at the decoder follows the same HMM-based speech synthesis as in [T98], first by deriving spectral parameters from the HMM and the transmitted state sequence (MLSA filters derived from the mel-cepstral coefficients of the state-sequence) and mixed excitation signal from a MELP decoder using band-pass voicing strengths associated with a feature vector of the HMM to enable increased naturalness. While no direct comparison with the previously established HMM segment vocoders of [T98] or [MTK98] is done, this method shows that an ergodic HMM with number of states as 128 yields a good overall quality and intelligibility with speaker characteristics preserved at an effective bit-rate of 128 bps, though no formal listening tests are done. The notions of the optimal number of states being 64 for American English to correspond to the phones in this work is based on earlier work [Pepp90, FC86, Pepp91] and is closely in corroboration with a related work on using ergodic-HMMs to model spoken languages for language-identification [RSS03, SR05].

2.4.4 Ismail and Ponting HMM Based Vocoders

In a work contemporary to that of [T98], Ismail and Ponting [IP97], Ismail [I98], and Ovens, Ponting and Turner [OPT00] propose a 300 bps HMM based vocoder also in a recognition-synthesis framework. This system is a variant of the methodology adapted in [T98] in at least 2 ways: Firstly, it uses a HMM that models abstract acoustic sub-word unit segments, rather than phones. In this context, it is also task-dependent and talker dependent and is also word-mediated in the sense that the HMM recognizer works with a vocabulary of 500 words, an associated pronunciation dictionary and acoustic models, and the recognized sub-word unit acoustic segment sequences are constrained to correspond to sequences of words from the known (500 word) vocabulary. Secondly, at the decoder, for synthesis it uses a parallel formant synthesizer, in tandem with a rule-based system that maps the recognized acoustic segments to formant parameters; i.e., it does not use HMM-based synthesis, quoting the trainable HMM synthesis framework of Donovan and Woodland [DW95] as computationally intensive for training for a single talker, though considering the system of [MTKI97] as viable for its rapid speaker adaptation ability.

In this system, the incoming speech is transcribed into a sequence of sub-word acoustic units with corresponding pitch and duration information. The sub-words are modelled around phones which are expanded to segments by defining the context and using a set of rules to define such an expansion. The HMMs modelled on such sub-word units allow for flexibility to extend the dictionary without having to train models afresh for new words. The pitch and voice information is also transmitted to add to the naturalness of the synthesized speech. This approach of HMM recognition followed by synthesis-by-rule where the synthesizer parameters are derived from natural speech results in a synthetic quality very nearly as the original speech. [OPT00] concludes that, though no formal evaluation of the system was performed, informal listener tests indicated that speaker-specific characteristics are preserved and that the re-synthesized speech sounds more natural than speech coded using a 2,400 bps LPC-based system.

2.4.5 Formant Trajectory Model Based Recognition-Synthesis

While the above sections dealt with various types of HMM based recognition-synthesis frameworks for ultra low bit-rate speech coding, we briefly describe the work of Wendy Holmes [H98] which is also set in the recognition-synthesis framework using HMMs, but with two important differences from the conventional HMM based recognition and synthesis, and being more in line with the work of Ismail and Ponting in Sect. 2.4.4 for the synthesis part: Firstly, the work in [H98] used a feature set comprising of the first three formant frequencies together with five mel-cestrum coefficients for a phone-like recognition using phone-level linear-trajectory segmental HMMs. The identified segments are represented by straight-line formant parameters for coding. Secondly, speech is synthesized at the receiver using a parallel-formant synthesizer (rather than HTS as in the systems described in Sects. 2.4.1, 2.4.2 and 2.4.3), driven by the frame-by-frame control parameters derived from the formant parameters.

The system was shown to produce speech with good intelligibility and preserving speaker characteristics at 600–1,000 bits/s. This work specifically highlights in being a unified model, in the sense of using the same linear formant-trajectory model for both recognition and system and further emphasizes that such a model used for recognition-based coding represents speech in such a manner that the model can be used for coding at a range of data rates, trading bits for a graded speech quality, incorporating increasing speaker characteristics at higher rates, over and above a baseline phoneme sequence synthesis (at the lower end of the data rate), though this paper focuses on the high bit-rate end of the range—i.e., coding formant trajectories—but not yet demonstrating the graceful degradations that the unified model is claimed to be capable of (for decreasing rates), which would have been a very appealing and dramatic phenomenological result of such a modeling approach, in the context of speech coding, and something yet to be attempted in most speech coding formulations.

2.5 ALISP Units and Refinements

In a significant departure from use of phone-like or diphone-like units to define the segmentation and labeling and the segment codebooks in most segment vocoder frameworks, a series of work reporting a coherent evolution of techniques first explored the use of ALISP units (Automatic Language Independent Speech Processing) [C98] that could span multiple phones as the coding units (for segmentation and labeling) and progressed towards different kind of recognition and synthesis units by refining the ALISP units in order to address various issues such as concatenation continuity, corpus based dynamic selection of units, speaker adaptation, robustness to noise, etc.. In the following we give an overview of this series of contributions, highlighting the main aspects of each publication even while maintaining the chronological progression of techniques proposed and evaluated.

2.5.1 Basic ALISP Framework

Primarily based on the early work of Cernocky [C98], the ALISP units were motivated towards finding alternative approaches to defining units in contrast to the conventional notions of sub-word units derived based on phonetic knowledge. The ALISP units are defined using a combination of techniques—temporal decomposition, unsupervised clustering (e.g. vector quantization) of the target vectors so derived and multigrams, followed by HMM based modeling of the segments associated with multigram labels. One of the application of the ALISP units in [C98] was for very low bit-rate speech coding, realizing intelligible speech at an average bit-rate of 120 bps for two sets of speaker-dependent experiments.

The earliest of the work reported in this direction is [CBC97b] where the authors follow their early experiences with multigrams and modified multigrams [CBC97a] similar to the earlier VVVQ formulation of [CL94]. The main part of this work is on the lines of proposing the techniques further reported in [CBC98a, CBC98b, B99] which is described in some detail below.

This procedure first applies a temporal decomposition technique of the speech feature vector sequence, first proposed by [A83], to derive a series of spectral events each consisting of a target and an interpolation function, representing speech as being made of steady-state vectors blended by the interpolation function as would typify an underlying articulatory process. The parameter vectors (e.g. LPCC vectors in [CBC98b]) located at the gravity centers of the interpolation functions are vector quantized to obtain a string of symbols, from which a set of characteristic variable length symbol patterns called multi-grams (MG) are derived, further leading to a dictionary of multi-grams; in [CBC98a], a MG dictionary of size 1,666 is used, made of 64 1-g, 1,514 2-g and 88 3-g, with corresponding average length of a sequence in terms of spectral events being 1.638 of 112.7 ms. Each sequence in this MG dictionary is represented by a HMM trained on a training corpus labeled by the MG entries. This yields HMMs that model variable length units of speech, but with an underlying correspondence with speech units defined by the multi-gram process, rendered meaningful by the fact that each such sequence is derived and quantized by a temporal decomposition. The HMMs were left-to-right, with number of emitting states being proportional to the number of temporal-decomposition events in the modeled sequence. The HMMs are used as in phone transcription, but now yielding variable length acoustic segments matching the sequences that each HMM models. The work reported natural synthetic speech, though with limited experiments, and pointed to various possibilities including different synthesis methods (PSOLA, MBROLA) and speaker adaptation.

Further, [CKBC99] combine the earlier reported approach of ALISP units (based on TD, VQ, and HMM) with syllable-segments which were derived from semi-automatic segmentation procedure used for a syllable speech synthesizer. The system was reported to operate at 175 bps on test data for ALISP derived units with a spectral distortion of 4.39 dB and having subjectively intelligible speech, though unnatural and with strongly audible artifacts. The syllable based approach had an average bit-rate of 62 bps with spectral distortion of 8.9 dB and subjectively poor quality of the synthesized speech which was hard to understand.

2.5.2 Re-segmented Long Synthesis Units

In a progression from the basic ALISP units, [CBC00] propose a technique at 120–195 bps with main emphasis on identifying the need to use a dichotomy of recognition (or coding) units (CU) and synthesis units (SU) and in going from phone-like units (defined from transition to transition, marked by intersections of the interpolation functions obtained from temporal decomposition) to long synthesis units that are defined from steady-state region to steady-state region, as marked by TD target vectors. Closely following this work, [MBCC01, MGC01] propose specific techniques for defining 'new units' that were derived from the 'original units' which were characterized by having unstable parts at their boundaries, contributing to 'transition noise' due to poor concatenation (high discontinuities). This paper further motivates the need to reduce these poor concatenation by proposing new units that can be obtained by a re-segmentation of the original units from stable part to stable part (steady-state to steady-state) so that the concatenation discontinuity is low. Specific techniques for such a re-segmentation is proposed, namely, according to middle frames of original units, or according to middle frames of middle states of original unit HMMs, or according to gravity centers of original TD-based units. This paper also marked a departure from LPC analysis-synthesis to HNM based synthesis considering the LPC framework to contribute to artifacts in the synthesized speech. This work also discusses the choice of 'synthesis units', for each coding unit wherein the best synthesis units (from three representatives) is chosen, using minimum DTW distance between a representative and an input speech segment, with the DTW path transmitted to the decoder. The paper reported a speaker dependent coder at 370 bps with a best spectral distortion of 5.5 dB.

2.5.3 Short Synthesis Units by Dynamic Unit Selection

[B02, BC03] mark a significant departure from the earlier ALISP units as well as the long synthesis units derived by re-segmentation towards the notion of short synthesis units with dynamic selection of units, close to corpus based text to speech synthesis, though not in the exact formulation of the classical unit-selection for TTS [HB96] or [LC01, LC02] to be discussed further in the subsequent chapters. This work makes a distinction between ALISP based recognition acoustic units or RAUs (as a codebook of HMM models trained on LPCC based ALISP units as in earlier work) and synthesis acoustic units (SAUs) with the speech segments representative of a synthesis unit being named synthesis speech representatives (SSR). With this distinction and definition in place, the dynamic unit selection in this work [BC03] is briefly outlined as below. Let the number of ALISP units used for input speech segmentation (also called recognition acoustic units or RAUs) be 64. Each class H_j is partitioned into 64 sub-classes, called H_iH_j containing all speech segments of H_j that have H_i as left-context. A subset of this sub-class (e.g. K segments) are used as synthesis representatives. In the coding phase, the ALISP based segmentation and labeling is derived for an input speech; this is followed by determination of the synthesis unit for each segment based on the H_iH_j information and transmits the index of the class and the synthesis representative for each segment. At the decoder, the synthesizer concatenates the synthesis units, in the process ensuring that the left-context continuity is maintained for each segment. While [B02] also addresses speaker independent coding and VQ based speaker clustering and adaptation in this framework, [BC03] reports an average bit-rate of 400 bps with good quality speech.

2.5.4 Pre-selection of Units

In a further refinement of the corpus based coding in [B03, PCB04a, PCB04b] propose an additional pre-selection of units according to F0 so that the final selection process is done by incorporating both prosodic and spectral information; the time-alignment between the segment to be encoded and the pre-selected synthesis is given as a linear length correction (rather than in the form of a precise alignment at the frame level through DTW). Different selection criteria such as correlation measure on pitch profile, energy profile and harmonic spectrum are proposed. The information necessary for retrieving the synthesis unit at the decoder (for an input segment) is the class index (coded with 6 bits for 64 classes and 64 HMM RAU models) and the unit index in the associated sub-class (4 bits for 16 closest units according to the averaged pitch). This work reports an average bit-rate of 481 bps (~500 bps) with intelligibility tests (DRT) scores of 80 % for the proposed coder, in comparison to 77 % for Stanag 4479 at 800 bps and 88 % for Stanag 4591 at 2,400 bps.

2.5.5 Noise Robustness

In the same framework as [B03, PCB04a, PCB04b] as above, [PCB05] study the effect of noise on HMMs and the selection of units. It considers various methods for noise robustness such as speech enhancement, joint feature extraction and noise reduction and noise robust modeling using parallel model combination. The paper focuses on unit recognition errors rather than synthesis unit selection, as this is considered robust due to inherent robustness of pitch (for SNR greater than 15 dB) and the closeness of synthesis units even in the presence of errors. Experimental results showed that AURORA MFCC features were more robust in terms of recognition errors when compared to MMSE MFCC and that PMC LPCC gave the best recognition errors.

2.6 Speaker Adaptation in Phonetic Vocoders

Within the basic framework of the phonetic vocoder in [PD89] (outlined in Sect. 2.2.3.5), most vocoders are based on a transcription and transmission of the input speech into similar phonetic units or some form of acoustic units along with prosodic information such as the duration of the units, average pitch (or pitch profile of the segments) and gain (or gain profile of the segments). As a result, they face a limitation that the input speaker characteristics are lost in the transmission and further synthesis at the decoder, i.e., all the transmitted information carries no (input) speaker-specific information (except perhaps the prosodic information, which when in the form of pitch and energy profiles as in some of the vocoder frameworks, can apparently convey some speaker characteristics). Moreover, the synthesized speech at the decoder is usually from the segment codebook in the form of templates or HMMs (of phone-like units) derived from some training speaker, and in the absence of any information about the input speaker at the decoder, the speech synthesized from the single-speaker trained synthesis codebook invariably carries the speaker identity of the training speaker. This makes such vocoder framework speaker-dependent (i.e., dependent on the speaker on which the synthesis unit models are trained).

In order to render such vocoders be able to synthesize speech at the decoder which is in close resemblance to the input speaker, it is necessary to employ special techniques that either incorporates the speaker information in the information extracted and transmitted at the encoder or uses some type of speaker adaptation technique to suitably adapt the segment codebook (obtained from a training speaker). This problem of speaker adaptation is addressed by Ribeiro and Trancoso in [RT96, RT97, RT98]. [RT96, RT97] propose an adaptation strategy involving transmitting the mean value and standard deviation of the radius and angle of the poles corresponding to formant frequencies for each phone. In the decoder, a speaker modification method alters the formant frequencies and bandwidths of vowel segments, by first retrieving a set of normalized values of each phone index, and restoring the RMS and LSP coefficients frame-by-frame, adapting the codeword to the input speaker by matching the mean value and standard deviation before duration normalization and speech synthesis using the adapted code words. [RT98] extend this work further to perform new codebook adaptation strategies, with gender dependence and interpolation frames, leading to better speaker recognizability and speech quality.

Following these contributions to speaker adaptation for phonetic vocoders, [RTC00] carry out intelligibility and speaker recognizability tests on their phonetic vocoders with and without speaker adaptation. The tests were based on the methodology used to select the new 2,400 bps DoD speech coder, where a listener is presented with a pair of utterances and is required to judge if they are spoken by the same or by different speakers. This work reports two sets of experiments: the first verified the degree to which each coder preserved speaker identity and the second, verified how well each coder preserved the information necessary to distinguish one speaker from another. Further tests included listeners rating the dissimilarity between two voices using a 6-point scale. This work also presented recent developments on their coder using the SpeechDat corpus of Portuguese that includes telephone calls from 5,000 speakers, and which allowed improvements to HMM models, codebooks and quantization tables and to study the performance with a wide speaker population.

2.7 Unit-Selection Paradigms

In what can be considered a very significant convergence of recognition, synthesis and coding, Lee and Cox [LC01, LC02] proposed a sub-1,000 bps coder which operated on the principles of 'unit selection' that is normally employed in text-to-speech synthesis using the concatenative synthesis methodology. Here, a large codebook (actually a continuous speech database) is used for selecting the appropriate segments that best match the input speech using a modified Viterbi decoding principle that incorporates the costs of both the segment quantization and the segment-to-segment continuity.

The first work [LC01] marks a major paradigm shift in segment-quantization for very low bit-rate speech coding, in the sense that the continuous codebook used is a 'single-frame codebook', i.e. a codebook of single frame vectors like a vector quantizer, but obtained without any clustering. The basic structure of this frame-level unit-selection segment quantization system of [LC01] is shown in Fig. 2.19. This is along the lines of the long vector codebook considered in the two reasonings in Sect. 1.3.5, which allows transparent quality quantization in the limit of large vector codebooks at rates 20 bits/frame and above. The Viterbi algorithm used for segment quantization performs 'unit selection' that favors quantizing consecutive frames of input speech using consecutive frames in the 'continuous codebook'; such an 'index contiguity' was further exploited using a run-length coding thereby

2.7 Unit-Selection Paradigms

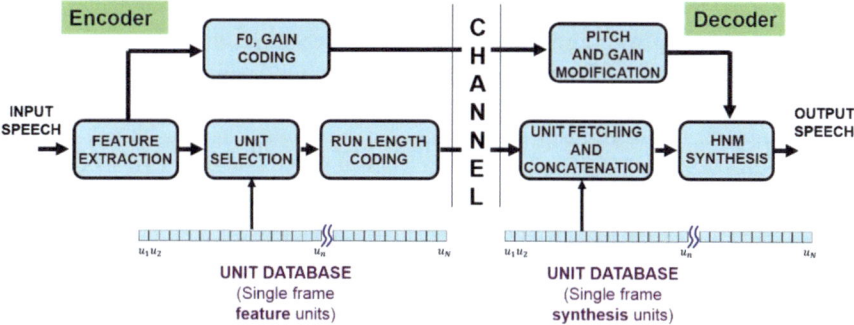

Fig. 2.19 Generic structure of the frame-level unit-selection based speech coding framework of [LC01]

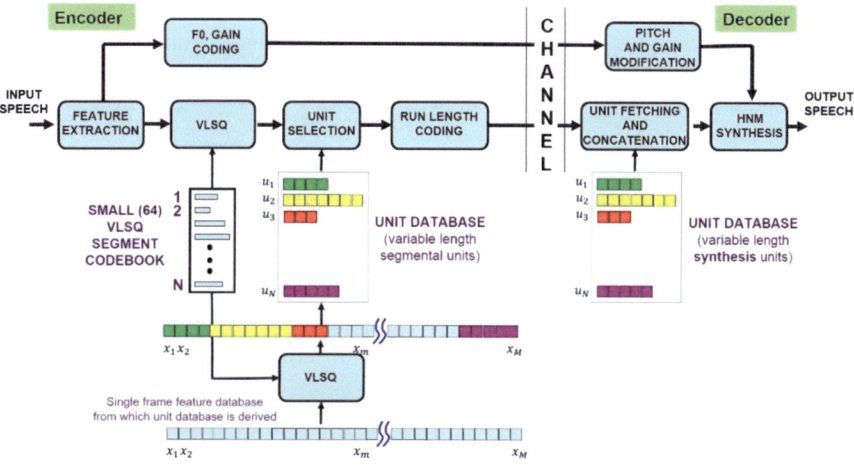

Fig. 2.20 Generic structure of segment-level unit-selection based speech coding framework of [LC02]

achieving low effective bit-rates though the codebook sizes used were significantly large (19 bits/frame). At the decoder, the received unit indices are used to retrieve the corresponding units from the unit database (now with a representation to aid synthesis) and concatenated prior to synthesis by a HNM synthesizer.

In another major milestone in large codebook based segment quantization, Lee and Cox also came up with a 'segmental' version of the above system [LC02] using a large size 'continuous codebook' (called 'unit database') made of variable length segments. Figure 2.20 shows the basic structure of this segmental unit-selection system of [LC02]. Here, Lee and Cox [LC02] used a segmental unit database for segment quantization of a 'pre-segmented and quantized' (by a VLSQ [SH88] quantizer) input speech, using a modified form of Viterbi decoding working on grouped units from the unit database and that favors contiguity of the quantization indices for run-length coding. At the decoder, the received unit indices are used to retrieve the corresponding units from the synthesis unit database and concatenated prior to synthesis by a HNM synthesizer.

In a recent and significant development in the unit-selection based segment quantization approach, we analyzed the algorithm of Lee and Cox [LC02], to show how it intrinsically suffers from several sub-optimality, such as due to pre-quantization of the test utterance using an intermediate Shiraki-Honda clustered segment codebook, and resulting fixing of unit labels and segment boundaries in test speech as well as the use of only a sub-set of units from the unit-database for concatenative unit-selection. Here [RH06, RH07], we proposed a unified and generalized framework for segment quantization of speech at ultra low bit-rates of 150 bps based on unit-selection principle using a modified one-pass dynamic programming algorithm and showed how it is exactly optimal for both fixed and variable-length segments and how it solves the sub-optimality of the Lee and Cox [LC02] algorithm by performing unit-selection based quantization 'directly' using the units of a continuous codebook without pre-quantizing the input speech. The unified unit-selection algorithm proposed by us handles both fixed- and variable-length units in a unified manner, thereby providing a generalization over both the two unit selection methods of Lee and Cox [LC01, LC02] which deal with 'single-frame' and 'segmental' units in a disparate manner. Moreover, the 'single-frame' algorithm of Lee and Cox [LC01] becomes a special case of the unified algorithm proposed by us.

In these algorithms presented by us [RH06, RH07], we showed three important results with respect to the early algorithms of Lee and Cox [LC01, LC02], based on rate-distortion curves using a very large continuous speech multi-speaker database: Firstly, our algorithm has a significantly superior performance than the segmental algorithm of Lee and Cox [LC02], by achieving considerably lower spectral distortions (up to 1.5 dB less) as well as much lower bit-rates over a range of database sizes; secondly, fixed length units perform significantly better than single-frame units, an important aspect overlooked by Lee and Cox's algorithm; and thirdly, use of fixed length units of 6–8 frames length offer spectral distortions similar to that of variable-length phonetic units, thereby circumventing expensive segmentation and labeling (manual or even automatic) of a continuous database (to define the variable-length units) for unit selection based low bit-rate coding.

Further, in [RH08], we proposed a low complexity unit-selection algorithm for ultra low bit-rate speech coding based on a first-stage n-best pre-quantization lattice and a second-stage run-length constrained Viterbi search to efficiently approximate the complete search space of the fully-optimal unit-selection algorithm recently proposed by us earlier [RH06, RH07]. By this, the proposed low complexity algorithm is rendered near-optimal in terms of rate-distortion performance while retaining the low complexity of the segmental unit-selection framework of Lee and Cox [LC02].

In another recent work [HR08], we analyzed the relative performances of different segment quantization methods, by considering five classes of segment quantizers used for low to ultra low rate speech coding, namely, vector quantization (VQ), matrix quantization (MQ), variable-length segment quantization (VLSQ) and the two unit-selection (US) based segment quantization algorithms US(LC) [LC01, LC02] and US(1-pass DP) [RH06, RH07] which represent an important shift in using large un-clustered continuous codebooks in contrast to the conventional clustered codebooks of VQ, MQ and VLSQ. Here, we examined the advantage, if any, in this shift from small clustered codebooks of VQ, MQ and VLSQ (10–14 bits/segment), to the larger continuous unit databases (16–18 bits/segment) in the unit-selection framework, by comparing their rate-distortion curves. We concluded in favor of variable-length segment quantization (VLSQ) over VQ and MQ and also showed how the unit-selection framework offers a promising direction in terms of providing steep fall in rate-distortion behavior.

In an important recent result [RH09], we showed that use of large unit databases (as in the above unit-selection framework for variable-length segment quantization) allows speech to be reconstructed at the decoder by using the best unit's residual itself (in the unit database), thereby obviating the need to transmit any side information about the residual of the input speech. For this, it becomes necessary to jointly quantize the spectral and residual information at the encoder during unit selection, and we proposed various composite measures for such a joint spectral-residual quantization within the unit-selection algorithm proposed earlier [LC01, LC02]. We realized ultra low bit-rate speaker-dependent speech coding at an overall rate of 250 bps using unit database sizes of 19 bits/unit (524,288 phone-like units or about 6 h of speech) with spectral distortions less than 2.5 dB that retains intelligibility, naturalness, prosody and speaker-identity.

More recently, in [R12], we extended this result further by considering the fact that the overall rate-distortion performance of such a joint spectral-residual quantization was compromised owing to the original sub-optimality of the unit-selection framework [LC02] within which it was set. In this work, in order to realize better rate-distortion performance, we proposed joint spectral-residual quantization in an optimal unit-selection framework based on the modified one-pass dynamic programming (DP) algorithm as in [RH06, RH07] By this, we realized ultra low bit-rate speaker-dependent speech coding with spectral distortions lower by up to 0.5 dB than the earlier algorithm [RH09].

The following chapters of this book (Chaps. 3 to 6) are primarily devoted to this unit-selection paradigm, specifically on the algorithms outlined in this section.

2.8 Performance Measures for Segment Quantization

The class of waveform coders can be evaluated with signal-to-quantization-noise ratio (SQNR) which measures the sample-to-sample distortion between the original speech waveform and the quantized and reconstructed waveform, by virtue of the fact that such coders preserve the sample-to-sample correspondence between the original and synthesized speech. However, in vocoders, starting from the LPC10 framework, and in the entire class of segment vocoders described in this chapter, such a sample-to-sample correspondence is not preserved due to the simple fact that the synthesizer (at the decoder) synthesizes speech using the quantized LP parameters and an approximation of the original residual derived from the prosodic parameters (voicing/unvoiced decision and pitch and gain) which renders the output speech waveform with no correspondence to the original speech waveform at the sample level.

In view of this, the LPC10 type of vocoders have been evaluated in terms of (i) the spectral quantization distortion, measured in terms of spectral distortion and (ii) subjective listening tests under various types of measures such as MOS, DRT, etc. In keeping with this practice, most segment vocoders and segment quantization techniques have adapted similar measures, among which the most preferred measure is the spectral distortion for the high-rate quantizers of LP spectral parameters [PK95]. We define this spectral distortion measure here, as this is employed in several segment vocoder work, and specifically in all of our work which we report in the following chapters (and outlined in Sect. 2.7).

The spectral distortion is a measure of the distortion between the spectra corresponding to the LP parameters of the original speech, as available at the encoder immediately after LP analysis (and prior to segment quantization) and the spectra corresponding to the LP parameters after segment quantization (and ideally, after channel transmission, though this is not relevant in the absence of any channel coding or errors) and as used in the LP synthesizer at the decoder, after duration modification, so that there is an one-to-one correspondence between the input frames (before quantization) and output frames (as used in the synthesis at the decoder). To this end, the two spectra between which the spectral distortion is computed corresponds to the points marked A and B in Fig. 2.5. The spectral distortion for a frame i is defined (in dB) as follows:

$$D_i = \sqrt{\frac{1}{F_s} \int_0^{F_s} \left[10\log_{10}(P_i(f)) - 10\log_{10}(\hat{P}_i(f)) \right]^2 df} \quad (2.4)$$

where F_s is the sampling frequency in Hz, and $P_i(f)$ and $\hat{P}_i(f)$ are the LPC power spectra of the ith frame given by

$$P_i(f) = \frac{1}{|A_i(j2\pi f/F_s)|^2}$$

and

$$\hat{P}_i(f) = \frac{1}{|\hat{A}_i(j2\pi f/F_s)|^2}$$

where $A_i(z)$ and $\hat{A}_i(z)$ are the original (unquantized) and quantized LPC polynomials, respectively, for the ith frame. Thus, the single frame spectral distortion is

2.8 Performance Measures for Segment Quantization

the squared difference between the log of the linear-prediction power spectra of the original frame and the quantized frame, averaged over frequency. The average spectral distortion is the average of the single frame spectral distortion given by Eq. 2.4, over the number of frames in the input speech. This average value represents the spectral distortion associated with a particular quantizer, and can be used as a quantization performance measure.

While we entirely rely on the above spectral distortion measure (SD in dB) to quantify and compare the performances of the different segment quantization schemes in the following chapters, several vocoders have also reported results of their segment quantization techniques using this measure, e.g., [KCT91, LF93, S94, CKBC99, MBCC01, C05]. Other measures that have been used include DRT [W82, RWR87, KCT91, PCB04a, PCB04b] and DAM [KCT91] and the PESQ measure [C08, CMC08b, PSVH10]. See also [TKCK93] for the kind of evaluation metrics and considerations that go into characterizing the performance of low rate coders in the range of 600–1200 bps, and which have a bearing on the evaluation of the coders operating at ultra low bit-rates too.

Note also that with the advent and promise of the unit-selection paradigm as demonstrated in the techniques outlined in Sect. 2.7, setting the basis for speech-to-speech synthesis in the unit-selection framework at ultra low bit-rates, it is also likely that objective and subjective measures used to characterize the quality of speech in the context of TTS (e.g. [vHvB95, D97]) could well be relevant for benchmarking the performance of such ultra low bit-rate coders.

Chapter 3
Unit Selection Framework

In what can be considered as a major paradigm shift in segment-quantization for very low bit-rate speech coding, Lee and Cox [LC01] proposed a system based on a unit-selection framework. This has several important distinctions from the conventional segment vocoder structure described in Chap. 2. Firstly, they used a 'continuous codebook', which is a sequence of mel-frequency cepstral coefficient (MFCC) vectors as obtained from continuous speech; the codebook is thus a 'single-frame codebook', i.e. a codebook of single frame vectors like a vector quantizer, but obtained without any clustering. Secondly, they employed a Viterbi decoding algorithm to perform segmentation and segment quantization using the 'unit selection' principle. Here, the Viterbi decoding uses concatenation costs which favor quantizing consecutive frames of input speech using consecutive frames in the 'continuous codebook'. The system then exploited this 'index-contiguity' to perform a run-length coding thereby achieving low effective bit-rates though the codebook sizes used were significantly large (19 bits/frame).

Subsequently, Lee and Cox also came up with a 'segmental' version of the above system [LC02]. Here, they used a similar large size 'continuous codebook' (called 'unit database', henceforth), but now segmented and quantized (i.e., labeled) by a 'clustered' codebook designed by the joint-segmentation quantization algorithm of Shiraki and Honda [SH88]. By this, the database now becomes a codebook of variable-length segments with each segment having an index from the clustered codebook. Lee and Cox [LC02] use this segmented and labeled database for a second stage quantization of the input speech, which is also segmented and quantized by the same clustered codebook. Here again, they apply a Viterbi decoding based unit selection procedure, but now to aid run-length coding on the unit indices of the database.

Figure 3.1 illustrates the principle of unit-selection based quantization for this segmental system of Lee and Cox [LC02]. The color sequence in the top shows the continuous speech which is segmented into variable length segments thus comprising the unit-database (each color bar representing a unit of some arbitrary length),

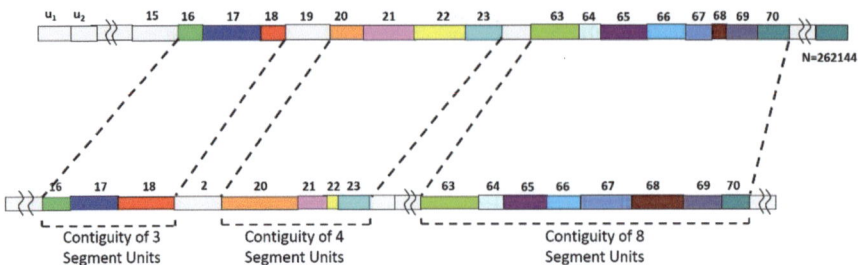

Fig. 3.1 Basic principle of unit-selection based segment quantization

say of length N = 262,144 units. The color sequence in the bottom represents the input speech being quantized by unit-selection principle. Here, the unit-selection based quantization yields a mapping between the input segments and the units, representing a quantization of an input segment by the corresponding unit. This solution is obtained in a manner identical to unit-selection based text to speech synthesis, wherein a total optimization cost (in this case, a distortion) is defined as a combination of unit costs and concatenation costs—the unit cost representing the distortion between an input segment and its corresponding quantizing unit and the concatenation cost defined in such a manner to favor units that are contiguous in the unit database to quantize consecutive segments in the input speech being quantized. Such unit contiguity is further exploited to yield low transmission bit-rates by employing run-length coding, where a contiguous sequence of units can be parsimoniously represented by the base index of the contiguous sequence of units and the number of units following the base unit that make up the contiguous sequence of units.

In this example shown in Fig. 3.1, three such contiguous groups are shown: three contiguous units 16, 17, 18 quantize three consecutive segments in the input speech, four contiguous units 20, 21, 22, 23 quantize four consecutive segments in the input speech and eight contiguous units 63–70 quantize eight consecutive segments in the input speech, each with its run-length advantage which is proportional to the length of the contiguity. This reveals the importance of the unit selection paradigm for quantization and the concatenation costs therein which facilitates the selection of such contiguous units. Note for instance, the input segment between the first group and the second group is quantized by unit number 2, thereby breaking the contiguity. Instead, if this segment had been quantized by unit 19, a long contiguous sequence of units 16–23 would have quantized eight consecutive segments in the input speech.

Note that this figure illustrates the basic principle of unit-selection based quantization where the unit database is made of variable length units. When the length of the units reduces to 1, this yields a single-frame continuous codebook as in the first algorithm of Lee and Cox [LC01], while the contiguity advantage holds good, but with a contiguous group of units being a sequence of single frames.

3.1 Lee-Cox Single-Frame Unit Selection Quantization

Let the input speech to be quantized using unit-selection be a sequence of T MFCC or LP vectors $O = (o_1, o_2, \ldots, o_t, \ldots, o_T)$. The unit database is a 'continuous codebook', which is essentially a sequence of MFCC or linear-prediction (LP) vectors as obtained from continuous speech. The unit database $\mathcal{U}$ is a sequence of N consecutive frames $\mathcal{U} = (u_1, u_2, \ldots, u_n, \ldots, u_N)$, where a unit u_n is a single frame vector.

It is desired to quantize the input speech $O = (o_1, o_2, \ldots, o_t, \ldots, o_T)$ by a sequence of units $U = (u_{q_1}, u_{q_2}, \ldots, u_{q_t}, \ldots, u_{q_T})$ from the database $\mathcal{U}$, in such a way to minimize a total quantization cost given by

$$D(O,U) = \sum_{t=1}^{T} D_u(o_t, u_{q_t}) + \sum_{t=2}^{T} D_c(q_{t-1}, q_t) \tag{3.1}$$

where, $D_u(o_t, u_{q_t})$ is the unit (or acoustic target) cost given by

$$D_u(o_t, u_{q_t}) = d(o_t, u_{q_t}) \tag{3.2}$$

where $d(o_t, u_{q_t})$ is the Euclidean distance between the MFCC vectors o_t and u_{q_t}. $D_c(q_{t-1}, q_t)$ is the concatenation cost given by

$$D_c(q_{t-1}, q_t) = \beta_{t-1,t} \cdot d(u_{q_{t-1}}, u_{q_t}) \tag{3.3}$$

Where, $d(u_{q_{t-1}}, u_{q_t})$ is the Euclidean distance between the units $u_{q_{t-1}}$ and u_{q_t}. $\beta_{t-1,t} = 0$ if $u_{q_{t-1}}$ and u_{q_t} are consecutive in the database and $\beta_{t-1,t} = 1$ otherwise. By this, the total cost is made to favor selection of $u_{q_{t-1}}$ and u_{q_t} that are contiguous in the database to quantize consecutive input vectors o_{t-1} and o_t. Such a concatenation cost attempts to preserve natural coarticulation and spectral continuity to the extent that a sequence of input speech can get quantized by a matching sequence of vectors in the unit database.

The optimal unit sequence $U^* = (u_{q_1^*}, u_{q_2^*}, \ldots, u_{q_{t-1}^*}, u_{q_t^*}, \ldots, u_{q_T^*})$ that quantizes the input $O = (o_1, o_2, \ldots, o_{t-1}, o_t, \ldots, o_T)$ with the minimum total distortion $D(O, U)$ is obtained as

$$U^* = \arg \min_{U \in \mathcal{U}} D(O, U) \tag{3.4}$$

where, U is any sequence of units, with each unit drawn from the unit database $\mathcal{U}$. Equation (3.4) is solved using a Viterbi search using the following recursions for the ith unit u_i in the database to be the unit that quantizes the input vector o_t at time t

$$D_t(i) = \min_{1 \leq j \leq N} \{D_{t-1}(j) + D_c(u_j, u_i)\} + D_u(o_t, u_i) \qquad (3.5)$$

$$\psi_t(i) = \arg \min_{1 \leq j \leq N} \{D_{t-1}(j) + D_c(u_j, u_i)\} \qquad (3.6)$$

This recursion is done time synchronously for $t = 1, \ldots, T$ and $i = 1, \ldots, N$ (at every t). $\psi_t(i)$ is the backtracking pointer for the ith unit at time t, $D_t(i)$ is the accumulated cost for the ith unit at time t and $D_c(u_j, u_i) = 0$ if the units u_j and u_i are consecutive in the unit database; i.e., $i = j + 1$.

The recursion is applied for $t = 1, \ldots, T$ and $i = 1, \ldots, N$ (at every t) and when $D_T(i)$, $i = 1, \ldots, N$ is computed, the optimal unit sequence $U^* = \left(u_{q_1^*}, u_{q_2^*}, \ldots, u_{q_{t-1}^*}, u_{q_t^*}, \ldots, u_{q_T^*}\right)$ is recovered from the backtracking pointer $\psi_t(i)$ using the following backward recursion

$$q_T^* = \arg \min_{1 \leq i \leq N} D_T(i) \qquad (3.7)$$

$$q_t^* = \psi_{t+1}(q_{t+1}^*), t = T - 1, T - 2, \ldots, 2, 1 \qquad (3.8)$$

3.1.1 An Alternate '5 ms Segment' Single-Frame Unit-Selection Algorithm

At this point, it is appropriate to refer to the work of [HT04] which sets out with the objective of showing how concatenative TTS with very short 'segment' durations of 5 ms is bound to be better than longer units, such as phones, diphones or syllables, mainly from the viewpoint that such short segments, in a given speech that will constitute the unit-database, provides more 'concatenation variation'—which is to be interpreted as offering more choice of such short segmental units in the database and which in turn have more concatenation options to match a given target specification derived from an input text. Though attempting to show this concept for TTS, this work resorts to 'speech-to-speech synthesis' (as discussed in Sect. 1.4), using short 5 ms segments for two reasons: (1) it is difficult to derive such short target vectors from text (for which the paper suggests use of a HMM based intermediate mapping to convert text into a sequence of target vectors which are to be further input to the proposed system) and, (2) it is easier to derive to such short 'acoustic vectors' directly from the speech signal (the paper uses fundamental frequency F_0, power and spectrum), and thereby be able to demonstrate that such short unit based concatenative synthesis does indeed produce acceptable quality speech, with the focus on the size of the unit.

Towards this objective, the paper actually realizes a 'speech coding' system incidentally, though without any accompanying standard objective quality measures or bit-rates that would normally qualify a coder. Instead, the paper does establish that short 5 ms units can yield natural speech, retaining the speaker identity (which is not surprising since both the unit-database speech and the test speech were from the same single speaker, and which would eventually result in the unit-database speaker to more naturally reflect in the synthesized speech, and which will invariably be mistaken as 'retaining' the same input speaker identity). More importantly, what is to be noted here is that the paper formulates the unit-selection step somewhat differently, starting with a n-best list of candidates from the unit-database for each 5 ms vector in the input, which is further searched for a sequence of units that have the best concatenation property, with appropriately defined concatenation-costs (e.g. KL distance on the normalized power spectrum at the concatenation points of the units).

We note here that the basic premise of this paper is questionable, firstly in not having provided fair comparison of different unit sizes in the same framework using objective measures, and secondly it runs contrary to the result shown by us in Sect. 4.1.2, where we have used a generalized framework for unit-selection with arbitrary sized units, and actually shown that the rate-distortion performances is enhanced considerably with unit-size, i.e., in progressing from single-frame units (as in [LC01] and in this paper [HR05]) to longer size units (e.g. 8 frames long) and further to variable-length phonetic units.

However, this work clearly belongs to the class of unit-selection based speech coding that is in focus here, though best seen as a (what seems independently arrived at by [HR05]) sub-optimal realization of the pioneering work of [LC01] which is the earliest single-frame unit-selection based speech coding as discussed above. The sub-optimality of [HR05] is due to the use of only the n-best candidates (a value of n = 300 was used in this paper) for each input vector, whereas it can be seen that the single-frame Lee and Cox [LC01] algorithm described above solves this by letting each input frame be quantized by (and concatenated to) any single-frame vector in the entire single-frame unit database.

3.2 Lee-Cox Segmental Unit Selection Quantization

Figure 3.2 gives a schematic of the 'segmental' unit-selection framework and algorithm used by Lee and Cox [LC02], described below:

1. **Unit database**: Consider a long continuous speech with continuous sequences of LP parameter vectors or mel-frequency cepstral coefficient (MFCC) vectors. A 'unit-database' is derived from this continuous speech by segmenting and quantizing (i.e., labeling) the continuous speech using a 'clustered' codebook $\mathcal{V} = \{v_1, v_2, \ldots, v_p, \ldots, v_P\}$ by variable-length segment quantization (VLSQ) [SH88]. This results in a unit-database which is a sequence of variable-length units $\mathcal{U} = (u_1, u_2, \ldots, u_n, \ldots, u_N)$, where a unit u_n is of length l_n frames, given by $u_n = (u_n(1), u_n(2), \ldots, u_n(l_n))$; each unit has an index (or label) from the

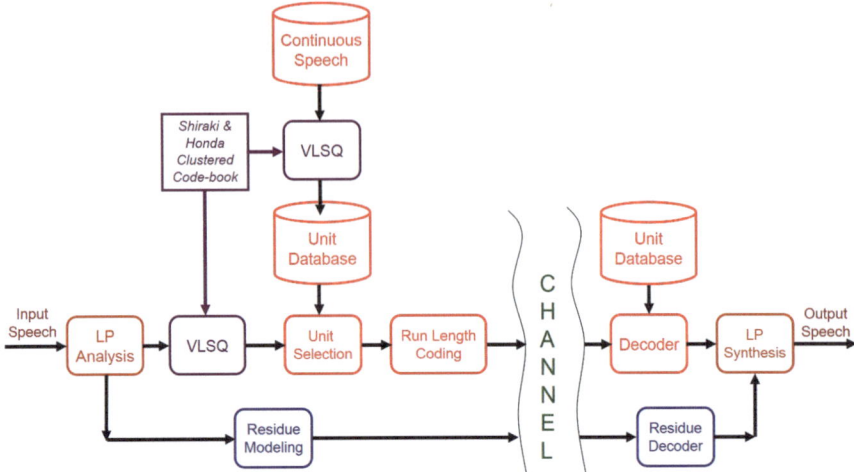

Fig. 3.2 Lee and Cox [LC02] segmental unit selection framework for variable length segment quantization (Reused with permission from [RH07])

clustered codebook $\mathcal{V}$, i.e., the label of unit u_n is $\mathcal{L}(u_n) \in [1,\ldots,P]$. Typical value of P is 64.

2. **Input speech segmentation**: The input speech (speech to be coded) is segmented and labeled by the same clustered codebook $\mathcal{V}$ using variable length segment quantization [SH88] (which is essentially realized by the 2-pass dynamic programming (DP) algorithm [RJ93]). If the input speech utterance is a sequence of vectors (MFCC or LP parameters) $O = (o_1, o_2, \ldots, o_t, \ldots, o_T)$, this results in a segmentation of O into K segments given by $S = (s_1, s_2, \ldots, s_{k-1}, s_k \ldots, s_K)$ with corresponding segment lengths $(L_1, L_2, \ldots, L_{k-1}, L_k \ldots, L_K)$. This segmentation is specified by the segment boundaries, such that the kth segment s_k is given by $s_k = (o_{b_{k-1}+1}, \ldots, o_{b_k})$. By virtue of the segmentation of O into K segments using the VLSQ codebook $\mathcal{V}$, each segment s_k is associated with a label from the VLSQ codebook $\mathcal{V}$, denoted by $\mathcal{L}(s_k)$ which can take a value from 1 to P.

3. **Unit Grouping**: Define a group $\mathcal{G}_k$ corresponding to each $s_k, k = 1, \ldots, K$ where $\mathcal{G}_k$ is a collection of all units in the unit database $\mathcal{U}$ such that the VLSQ label of all these units are the same as that of s_k, i.e., $\mathcal{G}_k = \{u_n : \mathcal{L}(u_n) = \mathcal{L}(s_k)\}$. This defines a collection of units $\mathcal{G}_k$ from the unit-database $\mathcal{U}$ for each segment s_k of the input speech utterance, which are now the potential candidate units for quantizing segment s_k.

4. **Unit selection segment quantization**: With the definition of $S = (s_1, s_2, \ldots, s_{k-1}, s_k \ldots, s_K)$ and $\mathcal{G}_k, k = 1, \ldots, K$ as above in Steps 2 and 3, it is now required to determine the optimal sequence of unit indices $Q^* = (q_1^*, q_2^*, \ldots, q_{k-1}^*, q_k^*, \ldots q_K^*)$ that minimize an overall decoding distortion (quantization error) when the segment sequence $S = (s_1, s_2, \ldots, s_{k-1}, s_k \ldots, s_K)$ is quantized by the corresponding unit sequence $\left(u_{q_k^*}, k = 1, \ldots, K\right)$. Before proceeding to give

3.2 Lee-Cox Segmental Unit Selection Quantization

the basic algorithm that solves this, we first give below the basic formalism that the above unit-selection quantization procedure solves.

The overall distortion (quantization error) in quantizing S by any Q is given by

$$D^* = \min_{Q} \left[\alpha \sum_{k=1}^{K} D_u(s_k, u_{q_k}) + (1-\alpha) \sum_{k=2}^{K} D_c(q_{k-1}, q_k) \right] \quad (3.9)$$

with the corresponding optimal unit sequence $Q^* = (q_1^*, q_2^*, \ldots, q_{k-1}^*, q_k^*, \ldots q_K^*)$ that quantizes $s_k, k = 1, \ldots, K$ being

$$Q^* = \arg\min_{Q} \left[\alpha \sum_{k=1}^{K} D_u(s_k, u_{q_k}) + (1-\alpha) \sum_{k=2}^{K} D_c(q_{k-1}, q_k) \right] \quad (3.10)$$

Here, $D_u(s_k, u_{q_k})$ is the unit-cost (or distortion) in quantizing segment s_k using unit u_{q_k} which is restricted to belong to $\mathcal{G}_k$. $D_c(q_{k-1}, q_k)$ is the concatenation-cost (or distortion) when unit $u_{q_{k-1}}$ is followed by unit u_{q_k}, but with the restriction that $u_{q_{k-1}}, u_{q_k}$ belong respectively to groups $\mathcal{G}_{k-1}$ and $\mathcal{G}_k$, i.e., $u_{q_{k-1}} \in \mathcal{G}_{k-1}$ and $u_{q_k} \in \mathcal{G}_k$. Further, the actual units $u_{q_{k-1}}, u_{q_k}$ selected from these groups are subject to the concatenation constraint

$$D_c(q_{k-1}, q_k) = \beta_{k-1,k} \cdot d(u_{q_{k-1}}(l_{q_{k-1}}), u_{q_k}(1)) \quad (3.11)$$

where, $d(u_{q_{k-1}}(l_{q_{k-1}}), u_{q_k}(1))$ is the Euclidean distance between the last frame of unit $u_{q_{k-1}}$ and the first frame of unit u_{q_k}. $\beta_{k-1,k} = 0$ if $q_k = q_{k-1}+1$ (i.e., $u_{q_{k-1}}$ and u_{q_k} are consecutive in the unit database) and $\beta_{k-1,k} = 1$ otherwise. This favors quantizing two consecutive segments (s_{k-1}, s_k) with two units which are consecutive in the codebook; run-length coding further exploits such 'contiguous' unit sequences to achieve lowered bit-rates.

A Viterbi decoding solves for the optimal sequence of unit indices $Q^* = (q_1^*, q_2^*, \ldots, q_{k-1}^*, q_k^*, \ldots q_K^*)$ that minimize the above defined overall decoding distortion (quantization error) D^* when the segment sequence $S = (s_1, s_2, \ldots, s_{k-1}, s_k \ldots, s_K)$ is quantized by the corresponding unit sequence $\left(u_{q_k^*}, k = 1, \ldots, K \right)$ under the constraints that

(a) Unit $u_{q_k^*}$ quantizes segment s_k and is drawn from the group of units $\mathcal{G}_k$,
(b) Two consecutive segments (s_{k-1}, s_k) are favored to be quantized by two units $u_{q_{k-1}^*}, u_{q_k^*}$ which are consecutive in the unit-database.

This is realized via a Viterbi decoding

(a) On a trellis of segment distortion values $\{D_u(s_k, u_{q_k})\}, k = 1, \ldots, K$ where $u_{q_k} \in \mathcal{G}_k$ and $D_u(s_k, u_{q_k})$ is an appropriately defined distortion between segment s_k and unit u_{q_k} and,

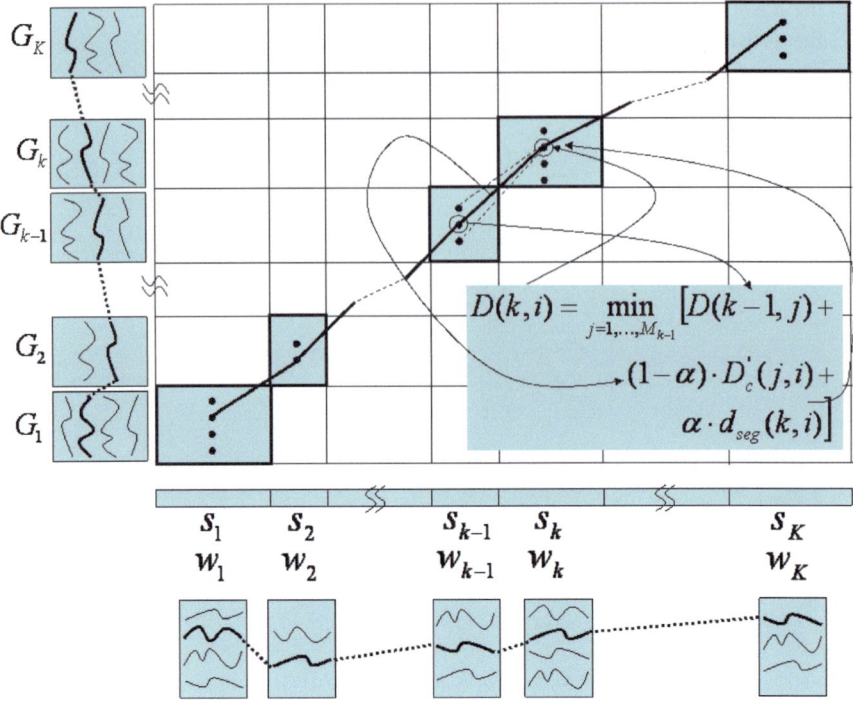

Fig. 3.3 Lee and Cox [LC02] segmental unit selection algorithm for variable length segment quantization

(b) By using concatenation costs which favor quantizing consecutive segments of input speech using consecutive units in the unit-database.

This is illustrated in Fig. 3.3. Specifically, we note the following:

(a) The x-axis shows the input speech $O = (o_1, o_2, \ldots, o_t, \ldots, o_T)$, segmented into K segments given by $S = (s_1, s_2, \ldots, s_{k-1}, s_k \ldots, s_K)$, by means of the variable-length segment quantization as described in Step 2 above.
(b) Following this, the y-axis is made of the K unit groups $\mathcal{G}_k, k = 1, \ldots, K$ each with M_k units drawn from the unit database, such that $\mathcal{G}_k = \{u_n : \mathcal{L}(u_n) = \mathcal{L}(s_k)\}$ as described in Step 3 above.
(c) The trellis based Viterbi decoding to find the optimal quantizing unit sequence indices $Q^* = (q_1^*, q_2^*, \ldots, q_{k-1}^*, q_k^*, \ldots q_K^*)$ is shown in the central part of the figure bound by the x- and the y-axes. As described in this Step 4, this involves finding the optimal path through the trellis that minimizes the total cost made of the unit-cost and the concatenation costs. The recursions employed in the Viterbi search for this optimal path is given as (as also shown in Fig. 3.3),

3.2 Lee-Cox Segmental Unit Selection Quantization

$$D(k,i) = \min_{j=1,\ldots,M_{k-1}} \left\{ D(k-1,j) + (1-\alpha) \cdot D'_c(j,i) \right\} \\ + \alpha \cdot d_{seg}(k,i) \tag{3.12}$$

where, $D(k,i)$ is the accumulated distortion of the path reaching segment s_k by quantizing it using a segmental unit $u_i \in \mathcal{G}_k$ (from among M_k possible units in this group). $D(k-1,j)$ is the accumulated distortion of the path at segment s_{k-1} having been quantized by a segmental unit $u_j \in \mathcal{G}_{k-1}$ (from among possible M_{k-1} units in this group). A path from $(k-1,j)$ to (k,i) has a concatenation cost of $D'_c(j,i)$ which takes on a value of 0 if u_j and u_i are contiguous in the original unit database, and takes on a value of the Euclidean distortion between the last frame of u_j and the first frame of u_i, if they are not contiguous. $d_{seg}(k,i)$ is the unit cost in quantizing segment s_k by unit u_i from the group $\mathcal{G}_k$ corresponding to the segment s_k. α defines the relative proportion of weighing the unit cost and the concatenation cost, as defined in the total cost definition in Eq. (3.9).

The recursion Eq. (3.12) is computed for $k=1,\ldots,K$, and for each k for all $i=1,\ldots M_k$. At the end of the recursions, all M_K path endings yield $D(K,i), i=1,\ldots,M_K$. Starting from the unit $q_K^* = \min_{i=1,\ldots,M_K} D(K,i)$, backtracking yields the best path and the best unit sequence $Q^* = (q_1^*, q_2^*, \ldots, q_{k-1}^*, q_k^*, \ldots q_K^*)$. This optimal solution by Viterbi search essentially selects the optimal unit $u_{q_k^*}$ from among the M_k units in group $\mathcal{G}_k$ to quantize input segment s_k, so as to minimize a global cost involving the unit costs of all the K segments and the unit-to-unit concatenation costs across segments. Such selected units in each group are shown in Fig. 3.3 by a bold segment below the x-axis under each segment s_k, and the dotted line connecting these bold segments is essentially the optimal path recovered by the Viterbi backtracking search.

This completes the segmental unit-selection for quantizing the input speech segments $S = (s_1, s_2, \ldots, s_{k-1}, s_k \ldots, s_K)$ by the optimal units $\left(u_{q_k^*}, k = 1, \ldots, K \right)$ from the unit database. Note that, frame level quantization of the input $O = (o_1, o_2, \ldots, o_t, \ldots, o_T)$ is further obtained by duration matching the unit $u_{q_k^*}$ to segment s_k by appropriate means.

5. **Run-length coding**: The system then exploited the 'index-contiguity' in the above unit-sequence Q^* to perform a run-length coding thereby achieving low effective bit-rates though the unit-database sizes used could be significantly large. Details of run-length coding and effective bit-rate calculations are given in Sect. 3.3.
6. **Duration modification**: Here, we add a note on 'duration modification' that is to be done at the decoder to make the unit u_{q_k} (of length l_{q_k}) to match the duration of the input segment s_k (of length L_k) that it quantizes, as well as in the computation of the unit cost $D_u(s_k, u_{q_k})$ in the Viterbi solution at the encoder above. This involves time warping of the unit u_{q_k} (of length l_{q_k}) to match the duration of the

input segment s_k (of length L_k). While Lee and Cox [LC02] employ a bi-linear interpolation to carry out this duration modification, we do this here by 'space-sampling'—a resampling of unit u_{q_k} to yield L_k frames so as to match the duration of s_k, which is as follows: The total length of unit u_{q_k} is computed (using a Euclidean norm on the LP vectors (say, LARs) of u_{q_k}; then the spectral trajectory of the unit u_{q_k} is resampled at L_k equispaced points in the LAR space. Thus, if the segment s_k is given by the sequence of frames $s_k(1), s_k(2), \ldots, s_k(l), \ldots, s_k(L_k)$ and the unit u_{q_k} is given by the sequence of frames $u_{q_k}(1), u_{q_k}(2), \ldots, u_{q_k}(l), \ldots u_{q_k}(l_{q_k})$, the resampled version of the unit u_{q_k} is given as $u'_{q_k} = u'_{q_k}(1), u'_{q_k}(2), \ldots, u'_{q_k}(l), \ldots u'_{q_k}(L_k)$, wherein the re-sampled unit u'_{q_k} is now matched in duration with the input segment s_k. The unit cost $D_u(s_k, u_{q_k})$ is then defined as

$$D_u(s_k, u_{q_k}) = \sum_{l=1}^{L_k} d\left(s_k(l), u'_{q_k}(l)\right) \quad (3.13)$$

3.3 Run-Length Coding and Effective Bit-Rate

Run length coding refers to the following coding scheme applied on the decoded label sequence obtained by the above algorithms. Let a partial sequence of labels in Q^* be $(\ldots, q^*_{i-1}, q^*_i, q^*_{i+1}, q^*_{i+2}, \ldots, q^*_{i+m-1}, q^*_{i+m}, \ldots)$ which are such that $q^*_{i-1} \neq q^*_i, q^*_{i+j} = q^*_i + j, j = 1, \ldots, m-1$ and $q^*_{i+m-1} \neq q^*_{i+m}$. The partial sequence $(q^*_i, q^*_{i+1}, q^*_{i+2}, \ldots, q^*_{i+m-1})$ is referred to as a 'contiguous group' with a 'contiguity' of m, i.e., a group of m segments whose labels are consecutive in the unit codebook. Run-length coding exploits this contiguity in coding the above contiguous group by transmitting the address of unit q^*_i first (henceforth referred to as the base-index), followed by the value $m - 1$ (quantized using an appropriate number of bits). At the decoder, this indicates that q^*_i is to be followed by its $m - 1$ successive units in the codebook, which the decoder retrieves for reconstruction. Naturally, all the m segment lengths $l_{i+j}, j = 1, \ldots, m-1$ are quantized and transmitted as in a normal segment vocoder.

Use of an appropriate concatenation cost favors the optimal label sequence to be 'contiguous' thereby aiding run-length coding and decreasing the bit-rate effectively. The unit-cost represents the spectral distortion and the concatenation cost (indirectly) the bit-rate; a trade-off between the two costs allows for obtaining different rate-distortion points for the above algorithm. This is achieved by the factor α (which takes values from 0 to 1).

The effective bit-rate with the run-length coding depends entirely on the specific contiguity pattern for a given data being quantized. For a given input utterance $O = (o_1, o_2, \ldots, o_T)$, let $Q^* = q^*_1, q^*_2, \ldots, q^*_{k-1}, q^*_k, \ldots, q^*_{K^*}$ be the optimal labels obtained by the algorithm as above. Let there be P 'contiguous groups' in this K-segment label sequence, given by $g_1, g_2, \ldots, g_p, \ldots, g_P$, where the group g_p has a

3.3 Run-Length Coding and Effective Bit-Rate

Continuous Codebook size = N =262144 No of bits/unit = log $_2$N = 18 bits		Maximum contiguity = C = 16 No of bits/unit = log $_2$C = 4 bits																Bits Calculation	Total Bits
Normal Sequential Coding		1 5	2 16	3 17	4 18	5 2	6 20	7 21	8 22	9 23								9 * log $_2$N = 9*18	162
Run Length Coding Case – I	Base Index	5		16		2		20										4log$_2$N = 4*18 = 72	88
	Contiguity	(0)		(2)		(0)		(3)										4log $_2$C = 4*4 = 16	
Run Length Coding Case – II	Base Index				63	64	65	66	67 63	68	69	70						1 * log $_2$N = 18	22
	Contiguity								(7)									1 * log $_2$C = 4	
Run Length Coding 16 Segments	Base Index	1	2	3	4	5	6	7	8 1	9	10	11	12	13	14	15	16	1 * log $_2$N = 18	22
	Contiguity								(15)									1 * log $_2$C = 4	

Fig. 3.4 Examples illustrating run-length coding principle and resultant bit-rate reduction advantage.

'contiguity' c_p, i.e., c_p segments whose labels are contiguous in the unit codebook. Then the total number of bits B for quantization of the input utterance O with run-length coding is given by

$$B = P \log_2 N + P \log_2 c_{max} + K^* \log_2 L_{max} \qquad (3.14)$$

where the first term is the total number of bits for the base-indices for the P contiguous groups, each being quantized to the address of the size N continuous codebook. The second term is the number of bits for the 'contiguity' information (providing for a maximum contiguity of c_{max} units) and the third term is the number of bits for the individual segment lengths in the K^* segment solution (providing for a maximum length of L_{max} frames). The effective bit-rate in bits/second is obtained by dividing this total number of bits B by the duration of the speech utterance Tf, for an input of T frames with a frame-size of f ms (20 ms in this paper with no overlap).

Figure 3.4 illustrates an example of the bit-rate advantage derived from such a run-length coding. Case-I corresponds to an input of 9 frames shown in Fig. 3.1 as quantized with two sections of contiguous units 16–18 and 20–23. Normal sequential coding (row 1), without using run-length scheme, quantizes each frame with 18 bits, with no contiguity advantage, yielding a net bit-rate of 162 bits for this sequence of frames. Case-I shows the net bit-rate due to unit-selection quantization with two run-length sections—16–18 with run-length of 2 and 20–23 with a run-length of 3, yielding a run-length based bit-rate of 88 bits. Case II shows the effective bit-rate for a longer contiguity—the section with units 63–70 in Fig. 3.1. This has one base-index (63) followed by the run-length value of 7, yielding an effective bit-rate of 22 bits—comprising of one base-index and one run-length value. Use of $\log_2 C = 4$ bits for quantizing the run-length contiguity allows representing a maximal contiguity of C, which in this case allows unit-selection of up to 16 contiguous units, as shown in the last row considering a sequence of units

1 to 16. The significant reduction in the effective bit-rate with high contiguity by means of unit-selection and run-length coding is the principle advantage of this scheme, as illustrated in this example.

3.4 Sub-optimality of Lee-Cox Segmental Unit-Selection Algorithm

In extending the 'single-frame' unit-selection principle to a 'segmental codebook', Lee and Cox [LC02] had introduced several sub-optimality in the segment quantization procedure, arising in the following ways:

1. Pre-quantization of the input speech before Viterbi decoding produces a segmentation that is sub-optimal with respect to the units of the actual units in the database.
2. The use of a unit-selection quantization on such pre-segmented and pre-quantized input utterance results in further loss of optimality as the optimal segmentation (and the corresponding quantization) would be significantly different, particularly with respect to the overall spectral distortion.
3. Using only those segments from the database which have the labels of the pre-quantized input speech restricts the units available for quantization to a small sub-set of units.
4. The unit selection Viterbi decoding essentially works only on segments defined by pre-quantization, and hence incurs a sub-optimality with respect to the overall spectral distortion of the final segmentation and quantization of the input speech with respect to the database units, which after all are the actual units used for synthesis at the receiver.

The focus of the following Chap. 4 is primarily on this sub-optimality issue and how a modified 1-pass dynamic programming (DP) algorithm proposed by us earlier [RH06, RH07], represents a unified and optimal algorithm for unit-selection based segment quantization, generalizing over both the algorithms of Lee and Cox [LC01, LC02], with consequent highly enhanced rate-distortion performances. Specifically, Sect. 4.1.1.1 provides a quantitative reasoning of the above sources of sub-optimality of the Lee and Cox [LC02] algorithm with reference to the optimal solution and search space of 1-pass DP algorithm to be discussed in detail in Chap. 4.

Chapter 4
Unified and Optimal Unit-Selection Framework

In this chapter, we present a unified framework for segmenting and quantizing the input speech using a constrained one-pass dynamic programming algorithm for performing unit-selection on continuous codebooks as used by Lee and Cox [LC01, LC02] for both single-frame and segmental unit-selection based quantization, which were described in detail in Chap. 3. Unlike the sub-optimal algorithm in [LC02], the algorithm proposed here provides an optimal segment quantization of the input speech with respect to the 'units' of the continuous codebook. Moreover, unlike the very disparate ways in which Lee and Cox realized the single-frame unit selection [LC01] and segmental unit selection [LC02], our proposed framework provides a 'unified' approach to treating the continuous codebook as made up of segmental units which can be of two kinds: (i) fixed lengths of arbitrary length (such as 1, 2, 3, 4, etc.) or, (ii) variable lengths such as phone-like units or units as derived after segmenting (and labeling) the continuous speech database using a 'clustered codebook' (as done in [LC02]). By this, we achieve several advantages over the methods of [LC01, LC02]:

1. The framework is based on a single elegant algorithm which is a generalization of both the single-frame system [LC01] and segmental system [LC02].
2. This allows evaluation of the unit-selection based system for fixed unit sizes greater than 1; this was not attempted in [LC01] or [LC02]. We show that using long fixed sized units of 6–8 frames offers significantly improved performance over 'single-frame' units [LC01].
3. We also show that using fixed size units of 6–8 frames (that approximate phone-like segments) offers performance comparable to variable sized units (such as phonetic units), thereby completely obviating the need to segment and label the continuous speech database manually or automatically using phonetic or clustered codebooks.

Figure 4.1 shows a schematic of the proposed unified and optimal unit-selection framework employing the modified one-pass DP algorithm for unit-selection based

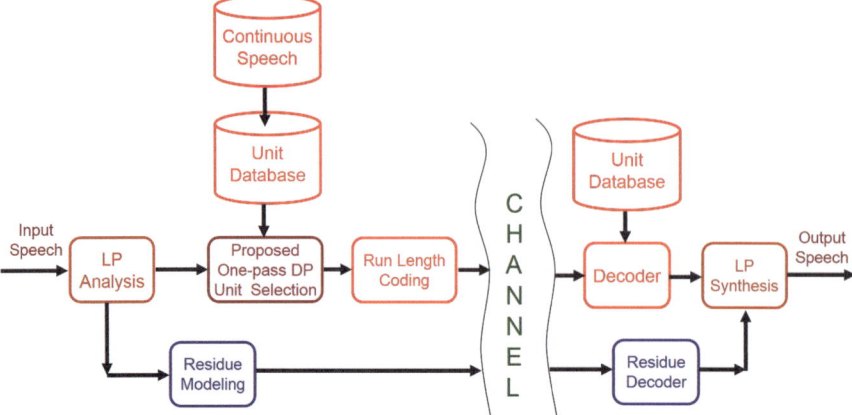

Fig. 4.1 Schematic of the proposed unified and optimal unit-selection framework employing a modified one-pass DP algorithm (Reused with permission from [RH07])

variable-length segment quantization directly on the units of the continuous unit-database. Note the absence of a first-stage variable-length 'pre-quantization' (using a clustered codebook) as employed in Lee and Cox [LC02] algorithm which labels the unit-database as well as the input speech.

This chapter is primarily based on the work reported in [RH06], [RH07] and [HR08], with several parts reproduced here with permission.

4.1 Unified Unit-Selection Framework

Consider a 'continuous codebook' $\mathcal{U} = (u_1, u_2, \ldots, u_N)$ which is essentially a sequence of MFCC or linear-prediction (LP) vectors as occurring in continuous speech, being composed of N variable length segments $(u_1, u_2, \ldots, u_n \ldots, u_N)$, where a unit u_n is of length l_n frames, given by $u_n = (u_n(1), u_n(2), \ldots, u_n(l_n))$. The codebook is said to be made of 'fixed length' units, if $l_n = l, \forall n = 1, \ldots, N$, i.e., each unit has l frames (when $l = 1$, the codebook is said to be a 'single-frame' codebook). The codebook is said to be made of 'variable length' units if l_n is variable over n.

Let the input speech utterance which is to be quantized using the above codebook be a sequence of vectors (MFCC or LP parameters) $O = (o_1, o_2, \ldots, o_t, \ldots, o_T)$. Segment quantization, in its most general form involves segmenting and labeling this sequence of vectors O by a 'decoding' or 'connected segment recognition' algorithm (as outlined in Sect. 2.2.2 and Sect. 2.2.3), which optimally segments the sequence and quantizes each segment by an appropriate label or index from the codebook. The segment indices and segment lengths together constitute the information to be transmitted to the decoder at the receiver, which then reconstructs a

4.1 Unified Unit-Selection Framework

sequence of vectors by concatenating the segments of the received indices after normalizing the original segments in the codebook to the received segment lengths.

In contrast, it should be noted that segment quantization based on 'unit-selection framework' (as dealt with in Chap. 3) differs from a conventional 'connected segment recognition' in minimizing a total distortion made of both the unit cost (or segment distortion) and concatenation cost (or unit-to-unit spectral discontinuity measure) which in turn makes the decoding solution to favor quantization of successive segments in the input by long contiguous units from the unit-database. In the following, we present an optimal formulation of this unit-selection based segment quantization which generalizes over and unifies the algorithms of [LC01] and [LC02] presented in Chap. 3.

Consider an arbitrary sequence of K segments $S = (s_1, s_2, \ldots, s_{k-1}, s_k \ldots, s_K)$ with corresponding segment lengths $(L_1, L_2, \ldots, L_{k-1}, L_k \ldots, L_K)$. This segmentation can be specified by the segment boundaries $B = ((b_0 = 0), b_1, b_2, \ldots, b_{k-1}, b_k \ldots, (b_K = T))$, such that the kth segment s_k is given by $s_k = (o_{b_{k-1}+1}, \ldots, o_{b_k})$. Let each segment be associated with a label from the codebook, with each index having a value from 1 to N; let this index sequence be $Q = (q_1, q_2, \ldots, q_{k-1}, q_k, \ldots, q_K)$.

Figure 4.2 gives the schematic of the basic formalism of the unified unit selection principle we use here. This generalizes to single-frame units, fixed-length units and variable length units (unlike the disparate formalisms adapted in Lee and Cox [LC01, LC02] for single-frame and variable-length segments).

The optimal decoding algorithm solves for K^*, B^*, Q^* so as to minimize an overall decoding distortion (quantization error) given by

$$D^* = \min_{K,B,Q} \left[\alpha \sum_{k=1}^{K} D_u(s_k, u_{q_k}) + (1-\alpha) \sum_{k=2}^{K} D_c(q_{k-1}, q_k) \right] \quad (4.1)$$

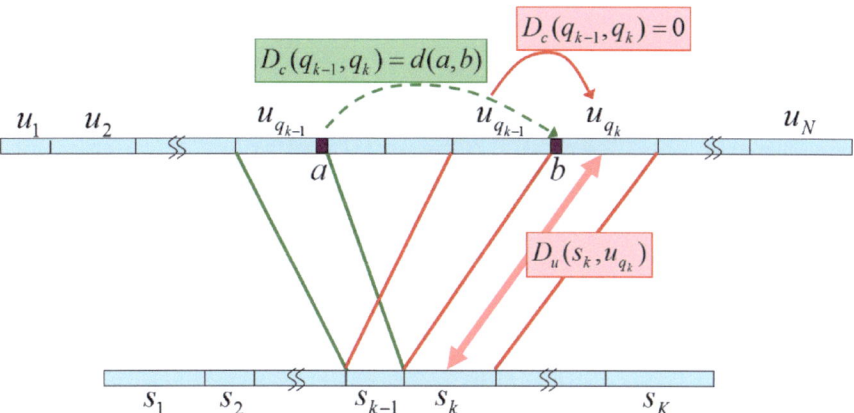

Fig. 4.2 Basic formalism of the unit-selection principle for generalized variable-length quantization

Here, $D_u(s_k, u_{q_k})$ is the unit-cost (or distortion) in quantizing segment s_k using unit u_{q_k}. This is as measured along the optimal warping path between s_k and u_{q_k} in the case of the one-pass DP based decoding which is described in Sect. 4.1.1. $D_c(q_{k-1}, q_k)$ is the concatenation-cost (or distortion) when unit $u_{q_{k-1}}$ is followed by unit u_{q_k}, which is given by

$$D_c(q_{k-1}, q_k) = \beta_{k-1,k} \cdot d(u_{q_{k-1}}, u_{q_k}) \qquad (4.2)$$

where, $d(u_{q_{k-1}}, u_{q_k})$ is the Euclidean distance between the last frame of unit $u_{q_{k-1}}$ and the first frame of unit u_{q_k}. $\beta_{k-1,k} = 0$ if $q_k = q_{k-1} + 1$ (i.e., $u_{q_{k-1}}$ and u_{q_k} are consecutive in the database and $\beta_{k-1,k} = 1$ otherwise. This favors quantizing two consecutive segments (s_{k-1}, s_k) with two units which are consecutive in the codebook, and more generally, favors several consecutive segments to be quantized by contiguous units in the unit-database; run-length coding (Sect. 3.3) further exploits such 'contiguous' unit sequences to achieve lowered bit-rates.

4.1.1 Proposed One-Pass DP Algorithm

We propose here a constrained one-pass dynamic-programming algorithm which performs an optimal segment quantization by employing 'concatenation costs' in order to constrain the resultant decoding by a measure of how 'good' is the sequence Q with respect to ease of run-length coding (described in Sect. 3.3).

We give here the modified one-pass dynamic programming algorithm to solve the above optimal decoding problem of Eq. (4.1). Figure 4.3 shows the structure of

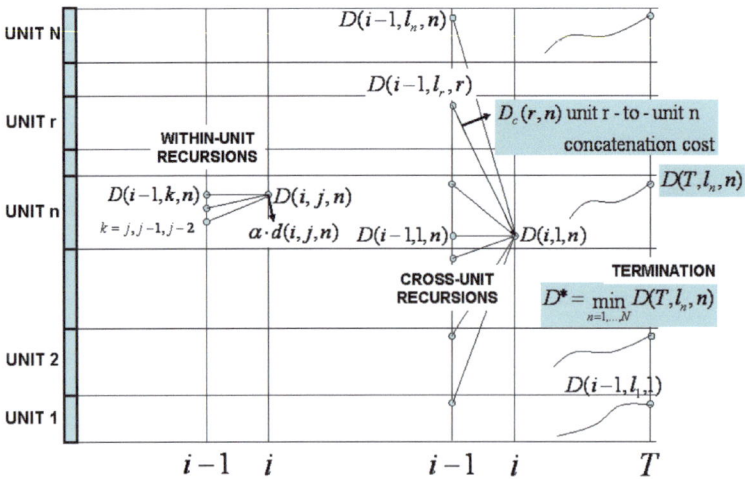

Fig. 4.3 Modified one-pass DP algorithm with the within-unit and cross-unit recursions to solve the basic unified and optimal unit-selection formalism

4.1 Unified Unit-Selection Framework

the modified one-pass DP algorithm, with the within-unit and cross-unit recursions, computed time-synchronously, using a unit database of N units (on the y-axis) to quantize the input speech of T frames (in the x-axis).

We state here the dynamic program recursions of our modified one-pass DP algorithm based unit-selection. The recursions are in two parts: within-unit recursion and cross-unit recursions.

Within-unit recursion

$$D(i,j,n) = \min_{k \in \{j, j-1, j-2\}} [D(i-1,k,n) + \alpha \cdot d(i,j,n)] \quad (4.3)$$

Cross-unit recursion

$$D(i,1,n) = \min(a,b) + \alpha \cdot d(i,1,n) \quad (4.4)$$

where,

$$a = D(i-1,1,n)$$
$$b = \min_{r \in (1,\ldots,N)} [D(i-1,l_r,r) + (1-\alpha) \cdot D_c(r,n)]$$

Here, the above two recursions are applied over all frames of all the units in the codebook for every frame i of the input utterance. The within-unit recursion is applied to all frames in a unit which are not the starting frame, i.e., for $j \neq 1$; the cross-unit recursion is applied only for the starting frames of all units, i.e., for $j = 1$, to account for a potential entry into unit n from the last frame l_r of any of the units $r = 1, \ldots, N$ in the codebook.

$D(i,j,n)$ is the minimum accumulated distortion by any path reaching the grid point defined by frame 'i' of the input utterance and frame 'j' of unit u_n in the codebook. $d(i,j,n)$ is the local distance between frame 'i' of the input utterance and frame 'j' of unit u_n. $D_c(r,n)$ is the concatenation cost (or distortion) when unit u_r is followed by unit u_n (accounting for the concatenation term $D_c(q_{k-1},q_k)$ in Eq. (4.1)) as defined in Eq. (4.2). α and $1 - \alpha$ respectively weigh the unit-cost and concatenation cost, thereby realizing Eq. (4.1) and providing a parameter for controlling the relative importance of the two costs in determining the optimal path (this is described further in the next section on run-length coding). The final optimal distortion is given by,

$$D^* = \min_{n=1,\ldots,N} D(T,l_n,n) \quad (4.5)$$

The optimal number of segments K^*, segment boundaries B^* and segment labels Q^* [corresponding to this optimal D^* in Eq. (4.1)] are retrieved by back-tracking as in the conventional one-pass DP algorithm [N84].

Figure 4.4 shows an example of the optimal quantization solution obtained by the proposed one-pass DP algorithm, giving the optimal path retrieved by back-tracking,

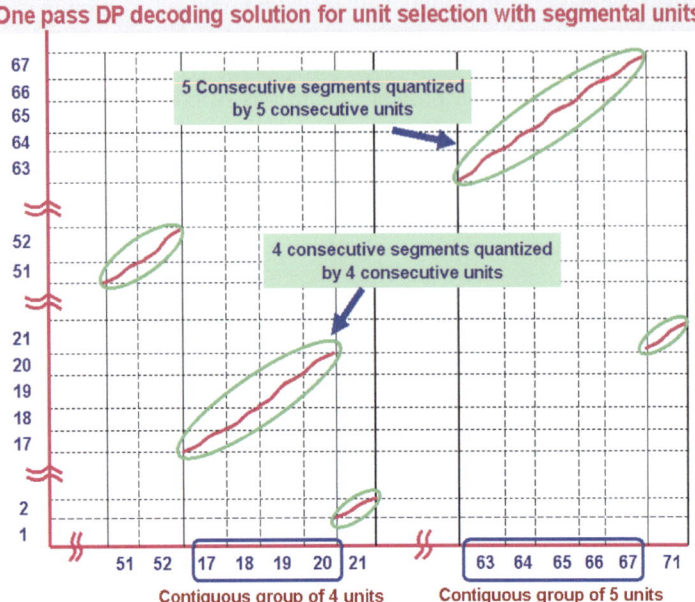

Fig. 4.4 Example of unit-selection solution obtained by the proposed optimal one-pass DP algorithm

which in turn is made of the quantization information and the unit-contiguity information. Note the three sections of contiguous units quantizing consecutive segments in the input speech—two segments quantized by contiguous units 51, 52; four segments quantized by four contiguous units 17–20; five consecutive segments quantized by five contiguous units 63–67.

The above modified one-pass DP algorithm can be viewed as a 'non-parametric' version of the Viterbi decoding employed in continuous speech recognition (CSR) using whole-word HMMs. The unit-selection framework here presents a unique setting for such a formulation, which incorporates a 'concatenation-cost' in the cross-unit recursions; (the conventional one-pass DP algorithm refers to this as 'cross-word transition', but does not use any transition cost). This concatenation cost corresponds to the word-to-word language model probabilities in the Viterbi decoding for CSR with word HMMs.

4.1.1.1 Comparison with Lee and Cox Single-Frame and Segmental Unit-Selection

The Viterbi algorithm used by Lee and Cox [LC01] with a 'single-frame' continuous codebook is a special case of the above one-pass DP algorithm when the units in the continuous codebook are of fixed length one. For variable length units, the above algorithm performs a decoding of the input utterance 'directly' using

4.1 Unified Unit-Selection Framework

the units of the unit codebook, unlike the two-stage procedure of Lee and Cox [LC02] which uses an intermediate segmentation (and labeling) using a clustered codebook (of size 64) followed by a conventional forced-alignment Viterbi decoding. As a result, we do not incur any of the sub-optimalities that the algorithm in [LC02] suffers from, as pointed in Sect. 3.4 qualitatively. This is made apparent quantitatively in the following.

Note that an optimal solution of unit-selection based quantization needs to solve Eq. (4.1) exactly, which is precisely what the modified 1-pass DP algorithm in Sect. 4.1.1. does. Stated differently, the optimal decoding needs to solve for (K^*, B^*, Q^*) that corresponds to the minimum quantization distortion D^*. While the optimal 1-pass DP algorithm solves this efficiently via dynamic programming, it indeed finds this optimal solution over the entire search space of (K, B, Q) considered jointly.

In contrast, the segmental unit-selection algorithm of Lee and Cox [LC02], as outlined in Sect. 3.2, incurs sub-optimality in its solution for the following reasons (as outlined qualitatively in Sect. 3.4):

i. The pre-segmentation and pre-quantization (by a joint segmentation and quantization) of the input speech using an intermediary clustered segment codebook $\mathcal{V}$ (in Step 2 of Sect. 3.2, say of size 64), solves for K' and B' first, where K' and B' are not the same as the globally optimal K^* and B^*, i.e., while K^* and B^* are optimal with respect to the 'units' of the unit-database $\mathcal{U}$, K' and B' are optimal only with respect to the intermediate clustered codebook $\mathcal{V}$, and are not optimal with respect to the unit-database $\mathcal{U}$ (whose units are after all the ones which are used in the synthesis at the decoder).

ii. The unit-selection in Step 4 (Sect. 3.2) solves for an optimal $Q' = (q'_1, q'_2, \ldots, q'_{k-1}, q'_k, \ldots, q'_{K'})$ in Eq. (3.10) by restricting $u_{q'_k}$ to be drawn from the group of units $\mathcal{G}_k$ to quantize the segment s_k (obtained by the 1st step pre-segmentation). The resultant Q' is clearly sub-optimal, as each of the segments s_k does not get to be quantized by the least distortion unit from the entire database $\mathcal{U}$, thereby leading to higher unit-costs $D_u(s_k, u_{q_k})$. Note that the concatenation cost $D_c(q_{k-1}, q_k)$ is also higher (than the optimal) owing to the fact that $u_{q_{k-1}}$ and u_{q_k} are restricted to be from the groups of units $\mathcal{G}_{k-1}$ and $\mathcal{G}_k$, thereby not allowing for the best possible concatenation of potential units from the entire unit-database $\mathcal{U}$.

Consequently, the solution (K', B', Q') obtained by the Lee and Cox [LC02] algorithm is sub-optimal in comparison to the globally optimal (K^*, B^*, Q^*) in the sense that the overall quantization distortion associated with (K', B', Q') with respect to the unit-database $\mathcal{U}$ is not the minimum possible and would be higher (in fact, 'significantly' higher, as will be shown in the results in Sect. 4.2) than D^* (in Eq. (4.1)) realized by the optimal 1-pass DP algorithm. The sub-optimality of Lee and Cox [LC02] segmental unit-selection algorithm can be viewed as essentially arising from the reduced search space that it considers for the choice of (K', B', Q'). As pointed out further in Sect. 5.4 (as part of a discussion on optimality vs complexity tradeoff), this reduced search space is what makes the Lee and Cox

[LC02] algorithm have a relatively low computational complexity in comparison to the fully optimal 1-pass DP unit-selection algorithm which finds its optimal solution over the entire search space.

Thus, the above 1-pass DP algorithm handles fixed-length segments of any size as well as variable length segments in a unified and optimal manner without taking recourse to two different ways of decoding as was done in [LC01, LC02] for single-frame and variable-length units respectively.

4.1.2 Experiments and Results

We now present results of the proposed unit-selection based segment quantization algorithm with respect to its quantization accuracy in terms of rate-distortion curves between spectral distortion and the effective bit-rate with run-length coding. We measure the segment quantization performance in terms of the average spectral distortion between the original sequence of linear-prediction vectors and the sequence obtained after segment quantization and length renormalization as defined in Sect. 2.8.

The average spectral distortion is the average of the single frame spectral distortion over the number of frames in the input speech; the single frame spectral distortion is the squared difference between the log of the linear-prediction power spectra of the original frame and the quantized frame, averaged over frequency. The bit-rate for segment quantization is measured as given in Eq. (3.14) in Sect. 3.3 using the run-length coding. We have used the TIMIT database for all the experiments. We have used a value of $\alpha = 0.5$ [Eqs. (4.1), (4.3), and (4.4)] in all the experiments, giving equal weightage to both unit-cost and concatenation cost.

In Fig. 4.5, we show the rate-distortion performance of the unit-selection algorithm for two kinds of unit sizes: (i) fixed length units with lengths ranging from 1 to 8 and (ii) variable-length phonetic units. In both cases, the codebook is a continuous sequence of linear-prediction vectors (log-area ratios) of continuous speech utterances in the TIMIT database, but treated as being made of fixed length units or variable sized units. Since TIMIT is phonetically segmented, we used this phonetic segmentation to define the variable-length units. This represents the best performance achievable for variable length units, such as when the automatic segmentation used to obtain the units is as good as a manual segmentation that defines phonetic segments. In both cases, we have used codebooks of sizes 32–4,096 which are essentially the first 32 (or 4,096) vectors of the TIMIT sentences ordered with male and female sentences interleaved. The numbers alongside each curve is the codebook size (in bits/unit). The number of sentences used to form these codebooks range from 1 to 128 sentences. The test data used was eight sentences with four male and four female speakers from outside the speakers used in the codebook.

From this figure, it can be observed that the effective bit-rate reduces significantly (nearly halves, such as from 200 bits/s to 100 bits/s), with increase in the fixed length unit-size from 1 to 4 to 8 frames. This is largely due to the fact that with a larger unit, the segment rate (number of segments per second) is reduced, and even without run-length coding, the number of bits used for base-index

4.1 Unified Unit-Selection Framework

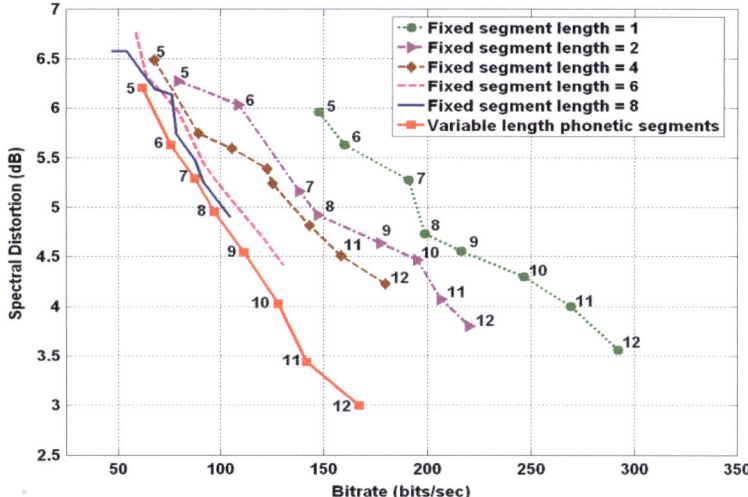

Fig. 4.5 Rate-distortion curves for different fixed length units and variable length phonetic units (Reused with permission from [RH06])

quantization would decrease proportionately. In addition, the use of run-length coding further reduces the effective bit-rate; the contiguity of larger length units implies that more frames are quantized with the same run-length base indices resulting in improved run-length advantage for longer fixed length units.

It can be noted that the variable length phonetic units performs the best, offering an halving of the bit-rate from 300 bits/s (for single-frame units) to 150 bits/s for the same distortion, clearly validating the potential of the unit-selection algorithm to gain rate-distortion with larger unit sizes that approximate phone-like units. However, fixed length units of length 6 and 8 (shown up to codebook sizes 4,096 and 2,048) also provide a performance comparable to that of variable length phonetic units. This circumvents the need for defining variable length units in a continuous codebook by automatically segmenting it or by other means. It would be sufficient to simply use a large continuous speech data and define fixed length units of lengths comparable to phonetic units.

The effect of increasing the fixed length on the run-length based bit-rate advantage is brought out clearly from the distribution of contiguity in Fig. 4.6 which plots the number of times a contiguity group of contiguity 'm' occurs. As can be expected, the contiguity is high even for units of length one. With increase in the unit lengths from 1 to 2, 4, 6, and 8, and finally to the variable length phonetic units, the largest contiguity tends to come down, since each unit now already spans multiple frames. However, the effective number of frames grouped by a contiguity has increased considerably even with limited contiguity for longer units. For instance, for unit lengths of 4, a contiguity of 4 performs an effective run-length coding over 16 frames in comparison to the maximum of 9 frames of single-frame units.

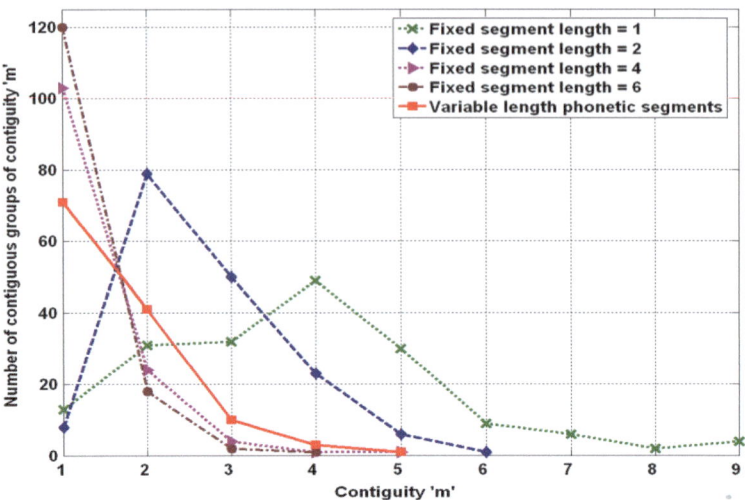

Fig. 4.6 Contiguity distribution for different fixed length units and variable length phonetic units (Reused with permission from [RH06])

It is important to note that, in Fig. 4.5, for longer fixed length units and variable length units, there is a sharper decrease in spectral distortion (SD) for a given bit-rate increase, in comparison to single-frame units. This steep trend in the rate-distortion curves of the proposed unit-selection algorithm even with fixed-length units, indicates that large reductions in spectral distortion can be achieved by using suitably large codebook sizes. This is particularly appealing since the codebook need not be 'designed' by any complex algorithms and nor does it have to be segmented (phonetically or otherwise) prior to use. Solutions to make the one-pass DP unit-selection algorithm perform with low computational complexities at very large codebook sizes, will enable it to achieve close to 1–2 dB average spectral distortions, which can make this unit-selection based ultra low bit rate quantization paradigm as good as high rate spectral quantizers.

4.2 Comparison with Lee and Cox Suboptimal Segmental Unit Selection

We now present results of the proposed unit-selection algorithm for segment quantization and compare it with the algorithm of Lee and Cox [LC02] in terms of quantization accuracy using rate-distortion curves between spectral distortion and the effective bit-rate with run-length coding, which are as defined in Sect. 4.1.2.

In Fig. 4.7, we show the rate-distortion performance of the unit-selection algorithm proposed by us here and the algorithm of Lee and Cox [LC02]. For both the

4.2 Comparison with Lee and Cox Suboptimal Segmental Unit Selection

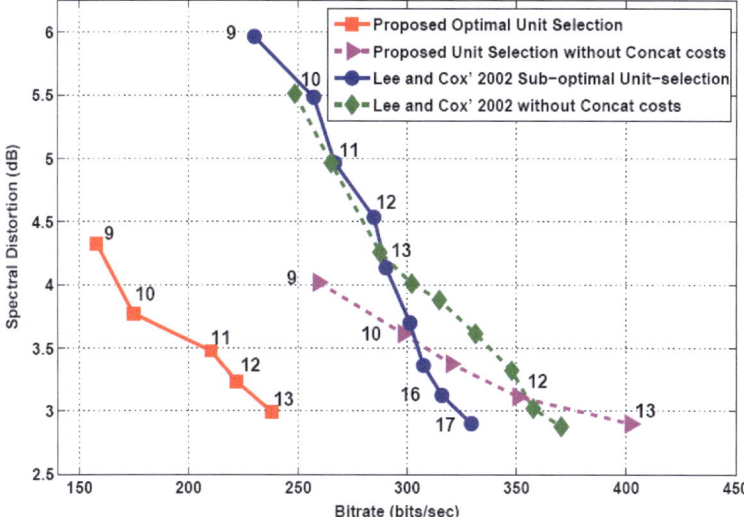

Fig. 4.7 Rate-distortion curves for proposed unified optimal unit-selection algorithm and the 2-stage suboptimal algorithm of Lee and Cox [LC02]. *Solid lines*: Unit selection with both unit-cost and concatenation cost (CC); *Dashed-lines*: Unit-selection without CC (Reused with permission from [RH07])

algorithms, we use the same continuous speech codebook as the 'unit database' which is a continuous sequence of linear-prediction vectors (log-area ratios) of continuous speech utterances in the TIMIT database, treated as being made of variable sized units as defined by the manually defined phonetic units. Since TIMIT is phonetically segmented, we have used this phonetic segmentation to define the variable-length units for both the algorithms. This represents the best performance achievable for variable length units, and can be expected to provide an optimal baseline performance to the case when automatic segmentation is used to obtain the units such as using a clustered codebook [by the variable-length segment quantization (VLSQ) technique] as used in Lee and Cox [LC02].

We have used 'unit databases' of size ranging from 512 to 65,536 corresponding bit-rates of 9 to 17 bits. These are essentially the first 65,536 phonetic segments of the TIMIT sentences ordered with male and female sentences interleaved, from about 200 sentences from 20 speakers constituting nearly 2 h of continuous speech. The number alongside each point in the curves is the codebook size (in bits/unit). In the case of the proposed algorithm, we have used database size up to 8,192, as this achieves the same spectral distortions as the Lee and Cox algorithm [LC02] with a database size of 65,536 and was hence adequate to bring out the performance advantage achievable (at significantly lower bit-rates), due to optimality of the proposed algorithm. The test data used was 10 sentences with 5 male and 5 female speakers from outside the speakers used in the codebook.

From this figure, the following important differences between the proposed optimal unified unit-selection algorithm and the sub-optimal algorithm of Lee and Cox [LC02] can be noted:

1. In general, the rate-distortion curve of the proposed algorithm has the ideal shift towards left-bottom with a significant distortion and rate margins over the rate-distortion curve of Lee and Cox [LC02]. This is as would be expected for an enhanced quantization scheme with both rate and distortion advantages.
2. Specifically, it can be seen that the proposed algorithm has significantly lower distortions for a given database size (given in bits alongside) and final effective bit-rate. For instance, for a database size of size 512 (9 bits), the spectral distortion of the proposed algorithm is about 1.5 dB less than that of Lee and Cox algorithm [LC02] with a corresponding effective bit-rate that is 75 bits/s less.
3. For a size of 13 bits, the proposed algorithm is able to provide a much lower spectral distortions (as much as 3 dB less) than the Lee and Cox [LC02] algorithm at the same effective bit-rate. It should be noted that this 3 dB difference is highly significant for the ultra low-rate ranges being dealt with here.
4. It can be further noted that the 13 bit database with the proposed algorithm gives about 1.5 dB performance improvement over that of Lee and Cox [LC02] and at a much lower bit-rate.
5. Further, it can be noted that the Lee and Cox [LC02] algorithm needs extremely large database sizes (of the order of 17 bits which is 65,536 segmental units or approximately 2 h of continuous speech), to achieve distortions comparable to that achievable by the proposed algorithm with a much smaller database size 13 bits (8,192 segmental units, or about 13 min of continuous speech, which is nearly 8 times less than that needed by Lee and Cox algorithm [LC02]).

We also show another important performance advantage of the proposed algorithm due to its optimality when compared to the sub-optimal algorithm of Lee and Cox [LC02]:

1. Figure 4.7 shows the rate-distortion curves of the two algorithms when the concatenation cost is not used; i.e., the Viterbi decoding in Lee and Cox [LC02] as well as the proposed one-pass DP constraints in this paper does not have the second term $\sum_{k=2}^{K} D_c(q_{k-1}, q_k)$ in Eq. (4.1). By this, the two algorithms have better (i.e., lower) spectral distortions, since not using the concatenation constraint leads to more optimal decoding with respect to the unit-cost of term 1 in this equation. It can be observed that the proposed algorithm has significantly lower spectral distortions than Lee and Cox [LC02] for a given unit database size. This clearly brings out the effect of gain of optimality resulting from quantizing the input utterance 'directly' using 'all' the units of the database, unlike the two step procedure of Lee and Cox [LC02] which uses an intermediate quantization (with a separate clustered codebook using the Shiraki and Honda VLSQ algorithm [SH88]) and a subsequent unit-selection using a

highly 'reduced' choice of units for each segment of the pre-segmented input utterance.
2. Further, it can be noted that the proposed algorithm gains significantly in achieving much lower effective bit-rates, once the concatenation-constraints are restored, again indicating another performance advantage of the optimality of the proposed algorithm: The proposed algorithm is able to produce more contiguous decoding, which in turn reduces the effective bit-rate with run-length coding. Again, this is due to the fact, the decoding is done with the entire unit-database, in comparison to the highly 'reduced' unit choices available in Lee and Cox [LC02] algorithm due to the pre-quantization using the clustered intermediate codebook.

While we have used phonetically defined variable length units as available in the TIMIT database, the above algorithms should in principle be used with a unit database defined automatically, i.e. with units defined by automatic methods such as the VLSQ method of Shiraki and Honda [SH88]. However, in an interesting result shown by us in an earlier work [RH06], it turns out it is possible to completely avoid such expensive segmentation and labeling (either manually or by automatic methods), by using fixed-length units of sufficient lengths (comparable to average phonetic units) such as 6–8 frames and still get rate-distortion performances comparable to what is possible with variable-length units. This leads to the conclusion that the 'optimal' algorithm proposed here is able to firstly overcome the sub-optimalities of the Lee and Cox [LC02] algorithm with a consequent improved rate-distortion performance and in addition, completely circumvent the need to have pre-defined variable-length units, as was obtained by using clustered codebooks in the Lee and Cox [LC02] algorithm.

4.3 Comparison with VQ, MQ, VLSQ

In this section, our objective is to benchmark the performances of some salient segment quantization algorithms using rate-distortion curves. This has more or less not been attempted at all, though [SH88] provides the early comparisons between VQ, MQ and VLSQ. However, here we put these early algorithms in perspective with respect to the recent unit-selection algorithms cited above. By this, our main intention is to bring out the important differences between the classical segment quantization schemes (VQ, MQ and VLSQ) and the current unit-selection based segment quantization algorithms, and provide insights into these differences and the causative factors. Primarily, as noted earlier, the difference comes about in that classical quantizers using clustered segment codebooks (fixed and variable length segments), whereas the unit-selection algorithms use large (long) continuous unit databases as in concatenative TTS. The question that arises is regarding what particular advantage does the use of very large continuous unit databases bring about (in the range of 16–18 bits/segment), in comparison to the much smaller clustered codebook sizes that VQ, MQ and VLSQ use (in the range of 8–10 bits/segment). Moreover, the early work of Lee and Cox did not also concern itself with quantifying the segment quantization performances in terms of rate-distortion curves, or answer the above question of what particular advantage has been gained by resorting to the unit-selection principles using large continuous codebook sizes. This section essentially attempts to answer this.

4.3.1 Experiments and Results

We now present results comparing the following segment quantizers, namely, vector quantization (VQ) [W82], matrix quantization (MQ) [TG85], variable-length segment quantization (VLSQ) [SH88] and the unit-selection algorithms of [LC02] discussed in Sect. 3.2 and [RH07] discussed in Sect. 4.1. The comparison is mainly in terms of quantization accuracy using rate-distortion curves between spectral distortion and the effective bit-rate (as appropriate in each case), which are as defined in Sect. 4.1.2.

In Fig. 4.8, we show the rate-distortion performance of these five quantizers (algorithms), obtained through different frame/matrix/segment codebook sizes for VQ, MQ and VLSQ and unit-database sizes for the unit-selection algorithms. For vector quantization, the VQ codebooks of size 16–2,048 (4–11 bits/frame) were

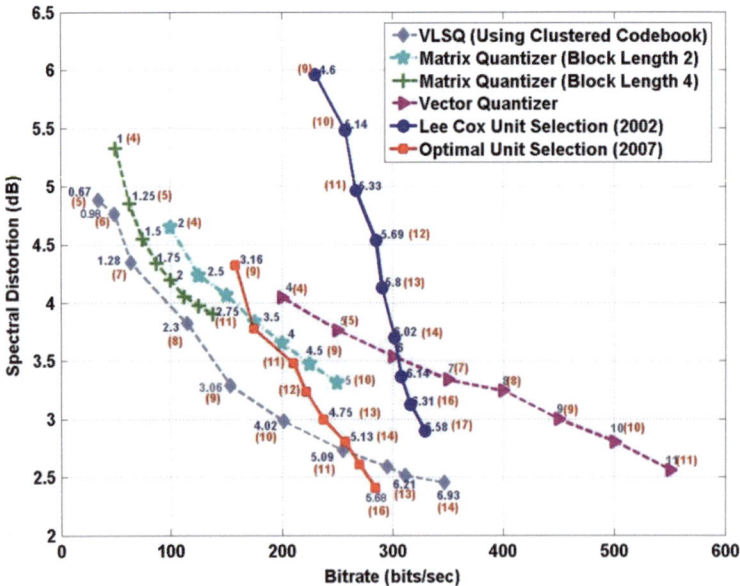

Fig. 4.8 Rate-distortion curves for VQ [W82], MQ [TG85], VLSQ [SH88] and the two unit-selection algorithms (i) Optimal algorithm [RH07] and (ii) Lee-Cox 2-stage algorithm [LC02] (Reused with permission from [HR08])

4.3 Comparison with VQ, MQ, VLSQ

designed from 48,000 frames of training data (320 sentences from 32 speakers, 16 male and 16 female) using the LBG algorithm and used for spectral quantization as given in [W82]. For matrix quantization, MQ codebooks of size 16–2,048 (4–11 bits/matrix) were designed for matrix block-sizes of 2 and 4 from the same training data as for VQ and used for quantization as in [TG85]. The VLSQ codebooks of size 32–16,384 (5–14 bits/segment) were designed by the joint segmentation and quantization algorithm of [SH88] from 90,000 frames of training data (600 sentences from 60 speakers, 30 male and 30 female) and used for segment quantization.

For both the unit-selection algorithms used here [LC02, RH07], we use the same continuous speech codebook as the 'unit database' which is a continuous sequence of linear-prediction vectors (log-area ratios) of continuous speech utterances in the TIMIT database, treated as being made of variable sized units as defined by the manually defined phonetic units.

Since TIMIT is phonetically segmented, we have used this phonetic segmentation to define the variable-length units for both the algorithms. This represents the best performance achievable for variable length units, and can be expected to provide an optimal baseline performance to the case when automatic segmentation is used to obtain the units such as using a clustered codebook (by the variable-length segment quantization (VLSQ) technique) as used in Lee and Cox [LC02].

We have used 'unit databases' of size ranging from 512 to 131,072 corresponding to bit-rates of 9 to 17 bits/unit. These are the first 131,072 phonetic segments of TIMIT sentences with male and female sentences interleaved, from ~200 sentences from 20 speakers of nearly 2 h of continuous speech.

The test data used for obtaining the R-D curves for all the quantizers was the same set of 8 sentences with 4 male and 4 female speakers from outside the speakers used in the codebook design for VQ, MQ and VLSQ and outside the unit-database for the unit-selection algorithms. In the rate-distortion curves in Fig. 4.8, the number alongside each point in the curves is the effective bits/frame (which is essentially the codebook size in bits/segment divided by the average length of a segment in the codebook, i.e., frames/segments); this yields the effective bit-rate in bits/sec when multiplied by the frame-rate of frames/sec, which in this case is 50 frames/s for a frame size of 20 ms). The numbers shown alongside each point within parenthesis is the codebook size (in bits/segment or bits/unit as appropriate). Both these are given to facilitate a quick comparison of the R-D performance of the different quantizers, either with respect to a given codebook size (which is appropriate when comparing VLSQ and unit-selection) or with respect to bits/frame which is more appropriate when comparing VQ, MQ and VLSQ, since these quantizers differ in the segment size in their codebooks. In the case of the proposed algorithm we have used database size up to 8,192, as this achieves the same spectral distortions as the Lee and Cox algorithm [LC02] with a database size of 65,536 and was hence adequate to bring out the performance advantage achievable (at significantly lower bit-rates), due to optimality of the proposed algorithm.

We observe the following from this set of R-D curves:

1. The VQ and MQ family of curves are as expected, with MQ of larger block sizes providing a left and downward shift the R-D curve; the reduction in spectral distortion (SD) for increase in block-size from 1 (VQ) to MQ(2) and MQ(4) for the same bits/frame is quite evident.
2. VLSQ offers improvement over VQ and MQ though only marginally with respect to MQ of block size 4.
3. When we shift to the unit-selection algorithms of Lee and Cox [LC02] or the optimal algorithm of [RH07], the codebook is unclustered and therefore these R-D curves have higher distortion for a given codebook size when compared to the clustered codebook performances of VLSQ, at least up to the maximum of 14 bit codebooks of VLSQ we have used.
4. However, the unit-selection algorithm reduce the spectral distortion more rapidly for every doubling of the codebook size, thanks largely to the run-length advantage of unit-selection and the associated reduction in the effective bit-rate which does not increase in proportion to the base-index bit-rate of the full codebook size. This results in a steep fall in spectral distortion even within 400 bits/s, while in contrast, VLSQ saturates at a SD of 2.5 dB for codebook sizes of size 16,384.
5. The advantage of the optimal unit-selection algorithm over the 2-stage sub-optimal segmental unit-selection of Lee and Cox can also be noted.
6. This optimal algorithm starts offering spectral distortions lower than VLSQ for considerably smaller unit database sizes than the sub-optimal unit-selection algorithm, and at a significantly smaller effective bits/frame than both VLSQ and the sub-optimal unit-selection.

In summary, we note that the unit-selection framework does offer an interesting rate-distortion trend of rapidly decreasing the spectral distortion for increase in the unit-database size, i.e., a steeper rate-distortion curve when compared to the VQ, MQ and VLSQ algorithms which tend to saturate in their spectral distortion reductions around codebook sizes of 10–14 bits/segment. This alone would be the distinctive factor that would allow unit-selection frameworks to offer distortion even in the range of 2 dB and less even though with use of very large continuous codebook sizes (perhaps exceeding even 18 bits/unit). More importantly, we believe issues related to computational complexity and memory and decoding latency time in the unit-selection algorithms will have to be addressed to take advantage of this rate-distortion trend and establish this class of segment quantizers as truly applicable for real ultra low bit-rate applications, in keeping with its seeming potential to offer low distortions with only marginal bit-rate increases, thanks to the run-length coding principles and advantages underlying the unit-selection framework.

4.4 Conclusions

We have proposed a unified framework for segment quantization of speech at ultra low bit-rates of 150 bits/s based on unit-selection principle using a constrained one-pass dynamic programming algorithm. The algorithm handles both fixed- and variable- length units in a unified manner, thereby providing a generalization over two existing unit selection methods, which deal with 'single-frame' and 'segmental' units in different ways. We show that fixed length units of 6–8 frames perform significantly better than single-frame units and offer the same spectral distortions as variable-length phonetic units, thereby circumventing expensive segmentation and labeling of a continuous database for unit selection based low bit-rate coding (Sect. 4.1).

We have brought out the intrinsic sub-optimalities of an algorithm proposed recently by Lee and Cox [LC02] for segmental unit-selection. We have proposed an alternative generalized unit-selection algorithm for segment quantization based on a modified one-pass dynamic programming algorithm. We have shown that the proposed algorithm is exactly optimal for variable length unit-selection based segment quantization and that it solves the sub-optimalities of the Lee and Cox [LC02] algorithm. Based on rate-distortion curves from a very large continuous speech multi-speaker database, we have shown that our algorithm has a significantly superior performance than the algorithm of Lee and Cox with considerably lower spectral distortions (up to 3 dB lower distortions) as well as much lower bit-rates for a given distortion over a range of unit database sizes (Sect. 4.2).

We have considered the class of segment quantizers used for low to ultra-low rate speech coding, ranging from vector quantization (VQ), matrix quantization (MQ), variable-length segment quantization (VLSQ) and two more recent unit-selection based segment quantization algorithms. We have examined the advantage, if any, in using large unclustered continuous unit databases by the unit-selection algorithms, in comparison to the smaller clustered codebook sizes that VQ, MQ and VLSQ use, by comparing the rate-distortion curves of these quantizers. We have shown that while unlike VQ, MQ and VLSQ, the unit-selection algorithms tend to provide lower distortions and steeper reductions at marginally low increase in bit-rates and justify exploring their potential further (Sect. 4.3).

Chapter 5
Optimality and Complexity Considerations

We propose a low complexity unit-selection algorithm for ultra low bit-rate speech coding based on a first-stage n-best pre-quantization lattice and a second-stage run-length constrained Viterbi search to efficiently approximate the complete search space of the fully-optimal 1-pass DP based unit-selection algorithm described in the previous chapter. By this, the n-best low complexity algorithm dealt with here, approaches near-optimality with increasing n, in terms of rate-distortion performance while having highly reduced complexity. The segmental unit-selection algorithm of Lee and Cox described in Chap. 3 is a 1-best special case of the algorithm proposed here, with the proposed n-best lattice based algorithm generalizing to a larger search space and hence a significantly improved rate-distortion performance towards the fully optimal 1-pass DP unit-selection performance.

In a recent work [RH06, RH07, HR08] (discussed in Chap. 4, Sect. 4.1), we proposed a unified and generalized framework for segment quantization of speech at ultra low bit-rates of 300 bits/s based on unit-selection principle using a modified one-pass dynamic programming algorithm. This one-pass DP algorithm offers the optimal rate-distortion performance, in the sense of the lowest spectral distortions with correspondingly low effective run-length based bit-rates that it accrues from the unit-selection framework. However, this algorithm suffers from a very high computational complexity of the order of $O(N^2T)$, where N is the number of units in the continuous unit database (of the order of $8-13 \times 10^5$ for a 17-bit unit-database) and T is the number of frames in the input test utterance being quantized (typically 50 for 1 s of speech). This represents a very high complexity for practical coding applications and it becomes important to find ways of reducing the complexity to practical proportions without sacrificing the optimal rate-distortion performance.

In this chapter, we elaborate on a recent result by us [RH08], where we proposed a n-best lattice based unit-selection towards rendering the sub-optimal Lee and Cox [LC02] algorithm more optimal, close to the fully optimal 1-pass DP unit-selection algorithm, even while retaining the low complexity that is inherent in the Lee and Cox [LC02] algorithm.

5.1 Complexity of 1-Pass DP Optimal Unit-Selection Algorithm

In Sect. 4.1.1, we presented the unified and optimal 1-pass DP algorithm for unit-selection based segment quantization. This algorithm is optimal in the sense of solving the segment quantization problem stated in the form of Eq. (4.1), and the 1-pass DP solution (in the form of recursions Eqs. (4.3) and (4.4)) for this problem realizes the minimum distortion D^* in Eq. (4.1). However, the 1-pass DP algorithm achieves this optimality at the cost of high computational complexity, which is the focus in this chapter. Specifically, we quantify this computational complexity incurred by the main two recursions [Eqs. (4.3) and (4.4)] primarily as a function of the unit-database size (N units).

We reproduce the recursions (4.3) and (4.4) here to show the inherent computational complexity arising from these. Figure 5.1 shows the details of this recursion and the associated complexity components.

Within-unit recursion

$$D(i,j,n) = \min_{k \in (j, j-1, j-2)} [D(i-1,k,n) + \alpha \cdot d(i,j,n)] \quad (5.1)$$

Cross-unit recursion

$$D(i,1,n) = \min(a,b) + \alpha \cdot d(i,1,n) \quad (5.2)$$

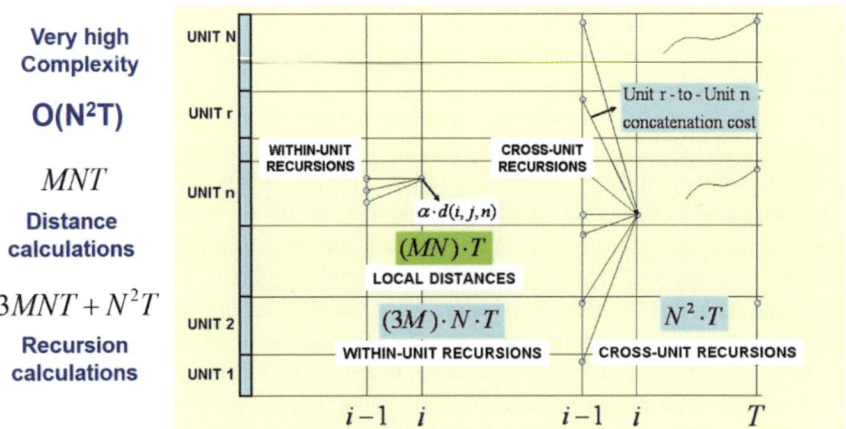

N – Number of units in continuous codebook → $2^{17} - 2^{20}$ units
M – Average number of frames per unit → 4-6 frames
T – Number of frames in input utterance → 50 frames for 1 sec

Fig. 5.1 Modified 1-pass DP recursion structure for unit-selection and associated complexity

5.1 Complexity of 1-Pass DP Optimal Unit-Selection Algorithm

where,

$$a = D(i-1, 1, n)$$

$$b = \min_{r \in (1,\ldots,N)} [D(i-1, l_r, r) + (1-\alpha) \cdot D_c(r, n)]$$

We give the computational cost in terms of three component costs (1) local distance calculations, (2) within-unit recursions, (3) cross-unit recursions.

5.1.1 Local Distance Calculations Cost

Both the within-unit recursion and the cross-unit recursion need $d(i, j, n)$, the local distance between frame i in the input test utterance and frame j of unit u_n in the unit-database. Given N units in the database, and assuming the average number of frames per unit to be M, this yields the dominant computational cost of MNT distance calculations, to compute $d(i, j, n)$, for $i = 1, \ldots, T$, $j = 1, \ldots, M$ and $n = 1, \ldots, N$.

5.1.2 Within-Unit Recursion Cost

The within-unit recursion in Eq. (5.1) is computed for $i = 2, \ldots, T$ and $j = 2, \ldots, M$ for each unit u_n, $n = 1, \ldots, N$ (excluding (a) the initialization for $i = 1$ for all frames in all units and (b) the cross-unit recursion for $i = 2, \ldots, T$ for $j = 1$ for all the N units). For each grid-point (i, j), $i = 2, \ldots, T$ and $j = 2, \ldots, M$, the recursion needs the min operator over 3 preceding grid points (at $i-1$), involving three comparisons. The total number of recursions for a frame i in the input utterance and unit u_n is $3(M-1)$ comparisons, yielding a total cost of $3(M-1)N(T-1)$ for all the N units in the unit-database and $(T-1)$ frames in the input utterance. This is $O(3MNT)$ comparisons.

5.1.3 Cross-Unit Recursion Cost

The cross-unit recursions in Eq. (5.2) is computed for $i = 2, \ldots, T$ for $j = 1$ for each unit u_n, $n = 1, \ldots, N$. For each i, the dominant computation in this recursion is the N cross-unit terms accounting for the concatenation cost $D_c(r, n)$, which is the cost for unit u_r to be followed by unit u_n, calculated as the Euclidean distance between the last frame of unit u_r and the first frame of unit u_n. While this is $O(N^2)$ distance calculations, this is independent of i, and can be calculated as a pre-processing step and stored with a storage complexity of $O(N^2)$. We therefore exclude this from the computational cost in the cross-unit recursions making up the quantization for a given input utterance.

Having excluded the fixed cost of $D_c(r,n)$, the other remaining computational cost recurring for each i are the N additions and comparison in the cross-unit recursions, to evaluate $\min_{r \in (1,...,N)} [D(i-1, l_r, r) + (1-\alpha) \cdot D_c(r,n)]$. This has a cost of N^2 per frame i of the input test utterance, and $N^2(T-1)$ for $i=2,\ldots,T$. We represent this as a cost of $O(N^2T)$ additions and comparisons. We note here that, though the cost of additions and comparisons in the above recursions are of significantly lower complexity than the distance calculations, we need to include this in the total complexity considering the large grid over which the calculations need to be made for large unit-database sizes N.

The total computational costs for the 1-pass DP optimal unit-selection algorithm are thus: $O(MNT)$ distance calculations, $O(3MNT)$ comparisons and $O(N^2T)$ additions and comparisons. Figure 5.1 gives typical numbers of M, N and T to give a measure of this total cost: $N = 2^{17}$ to 2^{20} units, $M = 4-6$ frames/unit and $T = 50$ frames for 1 s, (at a frame rate of 50 frames/s, using 20 ms framesize without overlap). We combine these as a total overall complexity of $O(MNT)$ distance calculations and comparisons and $O(N^2T)$ additions and comparisons, which we state as being made of two parts—$O(MNT)$ Euclidean distance calculations and $O(3MNT + N^2T)$ recursion calculations. Numerically, using $N = 2^{17}$, $M = 6$ and $T = 50$ for 1 s of input utterance, this translates to a cost of 2^{25} distance calculations and 2^{40} recursion calculations per second of input utterance. In order to highlight the quadratic dependence on N, we state the overall dominant complexity of the 1-pass DP algorithm as $O(N^2T)$, as shown in Fig. 5.1, considering its significant dominance over the $O(MNT)$ distance calculations, which is only linear in N.

As illustrated above, the overall complexity of the 1-pass DP optimal unit-selection is indeed very high for typical values of unit-database sizes and unit-lengths. Clearly, the optimality in terms of minimizing the overall quantization distortion and bit-rate comes at the price of a very high computational cost. We now obtain the computational cost of the Lee-Cox segmental unit-selection algorithm, and see that this algorithm, though sub-optimal in terms of the overall distortion and bit-rate, nevertheless has a very low complexity. We then view the issue of trade-off between optimality and complexity, and consider ways of rendering the Lee-Cox algorithm more optimal (i.e. closer to the optimality provided by the 1-pass DP algorithm), even while retaining the low complexity of the Lee-Cox segmental unit-selection framework.

5.2 Complexity of Lee-Cox Segmental Unit-Selection Algorithm

In Sect. 3.2, we described the Lee-Cox segmental unit-selection algorithm. The main computational part in this algorithm is as given in Eq. (3.12) which is the recursion employed in the Viterbi search for finding the optimal path that yields the solution to the segment quantization for an input utterance. Equation (3.12) is reproduced here for convenience of further discussion.

5.2 Complexity of Lee-Cox Segmental Unit-Selection Algorithm

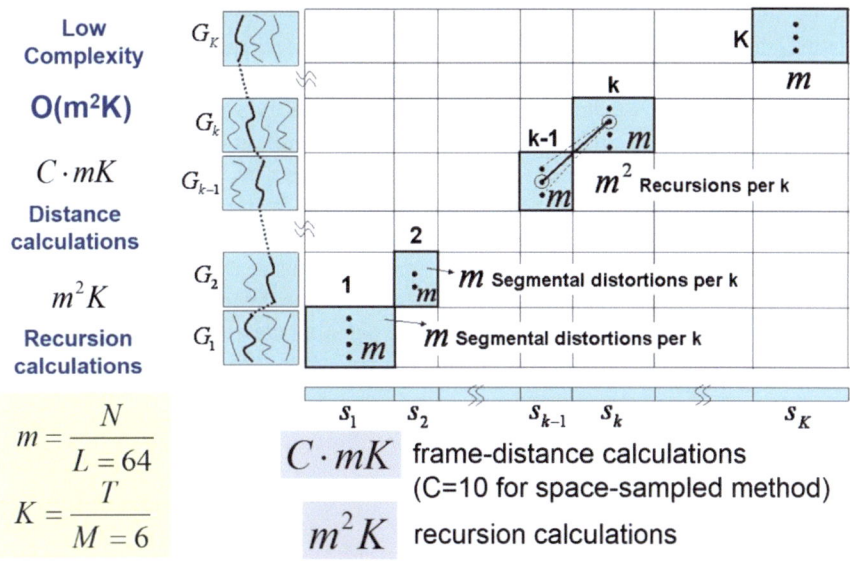

Fig. 5.2 Lee and Cox [LC02] segmental unit-selection algorithm, recursion structure and associated complexity

$$D(k,i) = \min_{j=1,\ldots,M_{k-1}} \left\{ D(k-1,j) + (1-\alpha) \cdot D'_c(j,i) \right\} \\ + \alpha \cdot d_{seg}(k,i) \tag{5.3}$$

This recursion corresponds to the trellis as given in Fig. 5.2 (which is essentially Fig. 3.3, but reproduced here highlighting the computational components). The trellis is made of K sections in the x-axis and K groups in the y-axis formed by the unit-grouping step (Step 3 in Sect. 3.2). In Sect. 3.2, we noted each group $\mathcal{G}_k$ to have M_k units. Here, for convenience of estimating the computational complexity, it can be reasonably assumed that each group $\mathcal{G}_k$ has the same number of units m; i.e., $M_k = m, k = 1, \ldots, K$. This corresponds to a roughly equal distribution (of $m = N/L$) of the original unit-database of N units into L disjoint groups, when the unit-database is derived by quantizing the original continuous speech database using a VLSQ codebook of size L (typically, 64 in the work here), as given in Step 1 of Sect. 3.2; i.e. the unit database of N units is divided into L groups each with $m = N/L$ units having the same index from the VLSQ codebook.

5.2.1 Distance Calculations

Figure 5.2 shows the trellis as having m entries (dots in the cell corresponding to section k of the input utterance, and the corresponding group $\mathcal{G}_k$), which are the m segmental distortions $d_{seg}(k, i)$ which in turn is the unit cost in quantizing segment

s_k by unit u_i from the group $\mathcal{G}_k$ corresponding to the segment s_k. From the assumption of m units in each group $\mathcal{G}_k, k = 1, \ldots, K$, the number of distance computations per k is m, totaling mK distance calculations for the entire input utterance. A single distance computation is actually a 'segmental' distance, between a segment s_k and an unit u_i in group $\mathcal{G}_k$. Ideally, a dynamic time warping based distance calculation between a segment s_k and unit u_i can have a complexity of $O(M^2)$ Euclidean distance calculations, where M is the typical length (in frames) of the two segments. However, using the space sampling method [RSM82b, RSM83] of computing this segmental distance, only a linear alignment is required, which in turn requires only C Euclidean calculations between the aligned segments, thereby making the total number of distance calculations as $O(CmK)$, with typical $C = 10$ [RSM83].

5.2.2 Recursion Calculations

The second computational component in Eq. (5.3), is the recursion from section $k-1$ to k, which requires m^2 additions and comparisons for $j = 1, \ldots, m$ and $i = 1, \ldots, m$ in Eq. (5.3), corresponding to m^2 transitions from each of the m entries in the cell at section $k-1$ to each of the m entries in the cell at section k. The total recursion cost for an input utterance of K sections involves applying the recursion (5.3) for $k = 2, \ldots, K$ and is $m^2(K-1)$, which is given as $O(m^2 K)$.

In summary, we state the computational costs of the Lee and Cox segmental unit-selection algorithm as $O(CmK)$ distance calculations and $O(m^2 K)$ recursion calculations, where the input utterance has K sections after a K-segment pre-quantization (Step 2 in Sect. 3.2), and $m = N/L$ for a unit-database size N and VLSQ codebook of size L. Typical values are $N = 2^{17}, K = 10$ for an input utterance of 1 second (or 50 frames), $L = 64$ and $C = 10$, yielding $O(2^{18})$ distance calculations and $O(2^{25})$ recursions calculations.

5.3 Comparison of 1-Pass DP and Lee-Cox Segmental Unit-Selection

It can be noted that Lee and Cox segmental unit-selection has several orders of magnitude less complexity than the optimal 1-pass DP unit-selection algorithm (Sect. 5.1). This comes from noting that $m = N/L$, and for $L = 64 = 2^6$, the $O(CmK)$ cost is 2^7 less than the corresponding cost of the 1-pass DP algorithm. This is easy to observe considering that each segment s_k in the input utterance is matched against only m units in a group G_k, and requires C distance calculations per segment, with the number of segments being K. This is in contrast to the 1-pass DP algorithm where a distance is calculated between every frame in the input utterance and all the frames of all the N units in the database.

Likewise, with respect to the recursion cost, Lee and Cox unit-selection algorithm has $O(m^2 K)$ cost, which, for $L = 64 = 2^6$ and $K = T/M$ for $M = 6$ is 2^{15} factor less than the corresponding cost of $O(N^2 T)$ of the 1-pass DP optimal unit-selection algorithm.

Noting that the quadratic complexity of both the algorithms is likely to be the dominating cost, it is clear that the optimal 1-pass DP algorithm is orders of magnitude more complex than the Lee and Cox segmental unit-selection algorithm.

5.4 Optimality-Complexity Tradeoff

Examining the two algorithms more carefully, the following can be noted with regard to how they trade off optimality and complexity.

1. The low computational complexity of Lee and Cox segmental unit-selection algorithm arises in primarily two ways: Firstly, the segmentation of the input utterance into K segments by a pre-quantization using the VLSQ codebook. By this, the problem reduces to a trellis of K sections, rather than all T frames as in the 1-pass DP optimal algorithm. This brings about a factor of T/M reduction in the trellis search complexity. Secondly, and more dominantly, the trellis involves only the units in each group $\mathcal{G}_k$ corresponding to segment s_k of the input utterance. Given that the number of units in a group $\mathcal{G}_k$ is $m = N/L$ (for $L = 64 = 2^6$), this yields a reduction of the search space to the highly reduced number of units in each group, as the candidates for the unit quantizing segment s_k.
2. On the same count, as we have noted earlier in Sect. 4.1.1.1, these reductions in the search-space is what leads to the sub-optimality of the Lee and Cox algorithm in terms of poorer rate-distortion performance.
3. In contrast, the fact that the optimal 1-pass DP algorithm makes no similar reductions in search space, and solves the unit-selection based segment quantization formulation of Eq. (3.1) exactly is what makes it optimal in the first place, and also makes it much more computation intensive than the sub-optimal Lee and Cox segmental algorithm.

From these observations, it immediately follows that we can either attempt to make the 1-pass DP algorithm less complex while retaining its optimality or to make the Lee and Cox segmental algorithm more optimal while retaining its low complexity. While both are indeed feasible, we focus on the latter, and discuss further a solution that renders the Lee and Cox segmental algorithm more optimal even while retaining its low complexity, in the process allowing means of reaching the optimality of the 1-pass DP algorithm at lowered complexity within the Lee and Cox unit-selection framework. To illustrate this, Fig. 5.3 shows the relative spaces occupied by the two algorithms with respect to the two aspects—complexity and optimality (measured in terms of spectral distortion). The 1-pass DP algorithm occupies the top-left region (marked A) characterized by low spectral distortions (high optimality) and high complexity. On the other hand, the Lee and Cox segmental algorithm occupies the right-bottom region (marked B), characterized

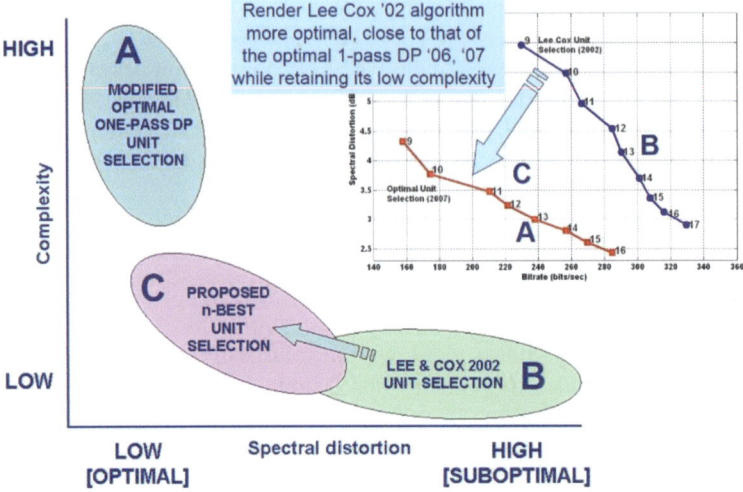

Fig. 5.3 Relative spaces (in optimality vs. complexity) occupied by the Lee and Cox [LC02] algorithm (B), optimal 1-pass DP unit-selection algorithm (A) and the proposed n-best lattice based unit-selection algorithm (C)

by high spectral distortions (low optimality) and low complexity. What we attempt here is to propose a means of rendering the Lee and Cox segmental algorithm more optimal, i.e., move the operating region more to the left (lower spectral distortions) (marked as C) possibly closer to the distortions of the optimal 1-pass DP algorithm, even while retaining its low complexity.

5.5 Proposed n-Best Lattice Search

In this section, we propose and discuss an n-best lattice search based two-stage unit-selection algorithm to render the Lee and Cox segmental unit-selection algorithm more optimal, close to the optimal performance of the more complex 1-pass DP algorithm, even while retaining the inherently low complexity of the 2-stage Lee and Cox segmental unit-selection framework.

The proposed n-best lattice search algorithm is set in the same framework as the Lee and Cox segmental unit-selection framework and differs from it primarily in the addition of a step involving defining the search space over which the trellis based Viterbi decoding works. In the following, we describe the algorithm with steps as in Sect. 3.2, but with the additional n-best lattice and corresponding expanded grouping and associated Viterbi search. The lattice search space, the trellis and Viterbi decoding are illustrated in Fig. 5.4.

1. **Unit-database and input segmentation**: The unit-database generation and input speech quantization are as in Step 1 and 2 of Sect. 3.2. Given the unit-

5.5 Proposed n-Best Lattice Search

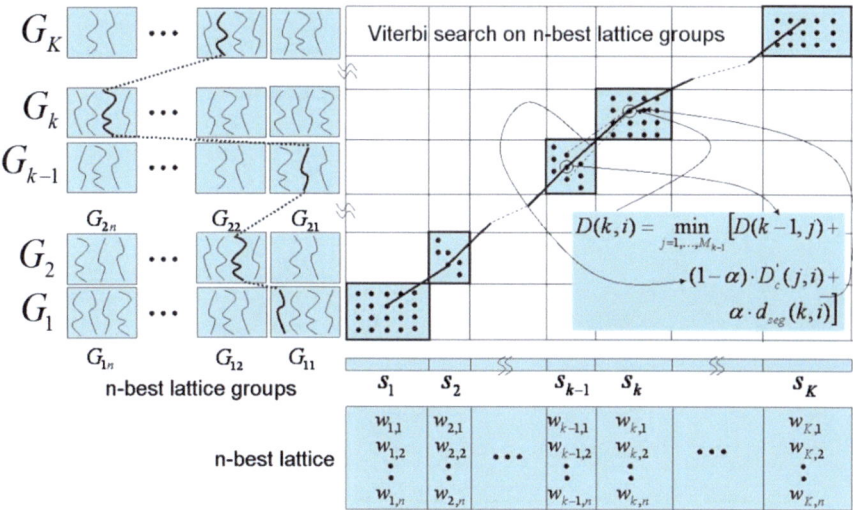

Fig. 5.4 Trellis and recursion structure of proposed n-best lattice based unit-selection set in Lee and Cox [LC02] framework

database, the input utterance $O = (o_1, o_2, \ldots, o_t, \ldots, o_T)$ is quantized using the variable-length segment quantizer (VLSQ) to yield a segmentation and corresponding VLSQ codeword labels.

2. **n-best lattice**: Following the segmentation of the input utterance into K segments, $(s_1, s_2, \ldots, s_{k-1}, s_k \ldots, s_K)$, derive a n-best lattice of this pre-quantization using the n-best VLSQ segment labels of each segment s_k, i.e., let the units w_{km}, $m = 1, \ldots, n$ be the n-best labels from the VLSQ codebook that have the least distortions with the input segment s_k.

3. **n-best grouping**: For each segment obtained after the above pre-quantization of the input utterance, hypothesize a collection of units from the large continuous unit-database having the same labels as the "union" of all labels in the n-best lattice for that segment s_k i.e., define group of units corresponding to segment s_k as $\mathcal{G}_k = \{\mathcal{G}_{k1} \cup \mathcal{G}_{k2} \cup \cdots \cup \mathcal{G}_{km} \cup \cdots \cup \mathcal{G}_{kn}\}$, where $\mathcal{G}_{km} = \{u_n : L(u_n) = w_{km}\}$ is the unit-group corresponding to the mth-best VLSQ label of segment s_k.

4. **Viterbi decoding**: Perform a constrained Viterbi-decoding with concatenation costs (to favor run-length sequences) on these (n-best lattice based) expanded unit-groups $\mathcal{G}_1, \mathcal{G}_2, \ldots, \mathcal{G}_k, \ldots, \mathcal{G}_K$.

The Viterbi decoding uses the same recursion Eq. (5.3), as in the original Lee and Cox unit-selection, but with the difference in the definition of the unit-group $\mathcal{G}_k$ as above. Each group $\mathcal{G}_k$ has approximately n times the original unit-group, being defined as the union of the units that have the n labels in the n-best list of segment s_k.

It can be seen that the Viterbi recursion now solves the basic unit-selection formulation of minimizing the total distortion given below [as given in Eq. (3.9)]

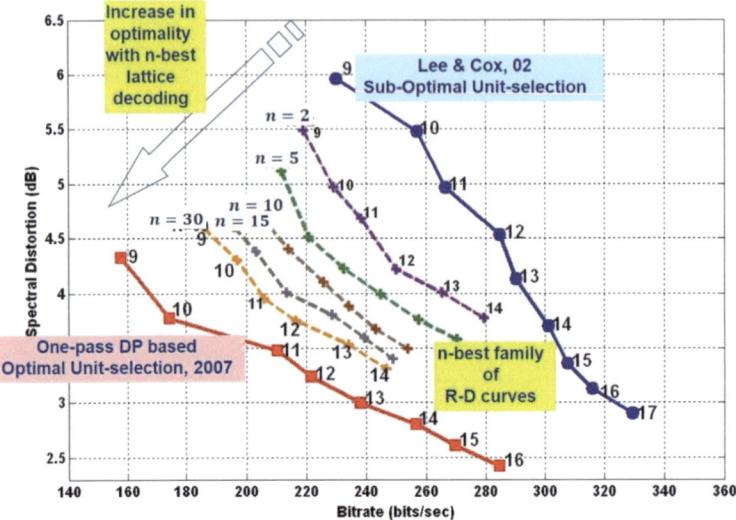

Fig. 5.5 Rate-distortion performance of proposed n-best lattice based unit-selection (family of curves for n = 2–30) with reference to the rate-distortion curves of the sub-optimal Lee and Cox [LC02] segmental unit-selection algorithm and the optimal 1-pass DP unit selection algorithm (Reused with permission from [RH08])

$$D^* = \min_{K,B,Q} \left[\alpha \sum_{k=1}^{K} D_u(s_k, u_{q_k}) + (1-\alpha) \sum_{k=2}^{K} D_c(q_{k-1}, q_k) \right] \quad (5.4)$$

The n-best lattice search algorithm given above solves for the optimal K^* and B^* from the first stage pre-quantization using the VLSQ codebook. However, the optimal sequence of units $Q^* = (q_1^*, q_2^*, \ldots, q_{k-1}^*, q_k^*, \ldots q_K^*)$ is obtained from the second stage Viterbi search of the n-best lattice groups, i.e., with the constraint that, in the solution to Eq. (5.4), the optimal unit $q_{k-1} \in \mathcal{G}_{k-1}$ and $q_k \in \mathcal{G}_k$, results in an increased search-space for both the choice of the best quantizing unit of s_k as well as in determining units q_{k-1} and q_k that are contiguous in the unit-database, which quantize consecutive segments s_{k-1} and s_k of the input utterance. Clearly, these two consequential result of the increased search space (by a factor of n, for a n-best lattice) can be expected to yield a better rate-distortion performance (i.e., lower distortions at lower bit-rates—lower distortions due to the availability of a larger search space in minimizing the unit-cost and lower bit-rates ensuing from better contiguity of units).

This improved performance with increase in n in the n-best lattice algorithm is clearly brought out in Fig. 5.5 which shows the overall spectral distortion realized at various overall bit-rates obtained for various unit-database sizes N. This figure shows the two base-line performances—that of the optimal 1-pass DP algorithm (red-line) and the sub-optimal (but low complexity) Lee and Cox [LC02] algorithm

5.5 Proposed n-Best Lattice Search

(blue-line). The family of rate-distortion curves for increasing n of the n-best lattice algorithm lies in between these two baseline rate-distortion curves. We make the following observations from this figure:

1. For $n=1$, the earlier unit-selection algorithm [LC02] becomes a 1-best special case (blue-circle, Fig. 5.1) of the n-best lattice algorithm proposed here; however, the proposed n-best hypothesis generalizes to a larger search space and has significantly improved rate-distortion (R-D) for increasing n (dashed line curves).
2. The proposed n-best lattice based algorithm offers a progressive lowering of R-D curves towards left-bottom for increasing n (from 1 to 30). Interestingly, even a doubling of the search space with $n=2$ helps in a significant jump in the rate-distortion curve (from the base-line Lee Cox [LC02] curve). Increasing n results in very pleasing reduction in the rate-distortion curve and even $n=30$ reaches a performance that is 'near-optimal' to fully-optimal one-pass DP algorithm (the lowest and best R-D curve at the left-bottom—red-square). Further increase in n will certainly approach the performance of the 1-pass DP algorithm, but more gradually, and more importantly, with commensurate increase in the complexity of the n-best lattice, which quickly reaches that of the 1-pass DP algorithm's complexity for higher n.

The primary observation here is to note that the Lee and Cox algorithm has been rendered to realize better rate-distortion performance by the n-best lattice based generalization for unit-selection, and very tangible performance improvements are indeed possible over the sub-optimal rate-distortion performance of the Lee and Cox [LC02] algorithm, even while retaining its low complexity (as will be shown in the following results and discussions).

It now remains to be seen whether the expected improvement in rate-distortion performance of the above n-best lattice based search is realized at the cost of increased complexity due to the use of the n-best lattice grouping in the Viterbi search. Following the same line of complexity derivation for the basic Lee and Cox segmental unit-selection algorithm, it can be seen that the complexity of the proposed n-best lattice has a multiplicative constant n to both the distance calculation cost $O(CmK)$ and the recursion costs $O(m^2K)$ of the Lee and Cox basic algorithm arising from the fact that each n-best lattice group has n times the number of units in group $\mathcal{G}_k$ when compared to the basic Lee and Cox unit-groups, i.e., the n-best lattice has $O(CnmK)$ distance calculation cost and $O(Cn^2m^2K)$ recursion costs. This is illustrated in Fig. 5.6.

In Table 5.1, we summarize the complexities of the three different algorithms considered here, namely, (i) optimal 1-pass DP unit-selection, (ii) sub-optimal Lee and Cox [LC02] segmental unit-selection and (iii) optimality-enhanced n-best lattice based unit-selection. Shown are the complexities of distance calculations and recursion calculations and the overall complexity considering the quadratic recursion complexity to be the dominant one. The last column shows the equivalence of the overall complexities, providing a direct comparison of the three algorithms, starting with that of the 1-pass DP as the reference ($O(N^2T)$), further

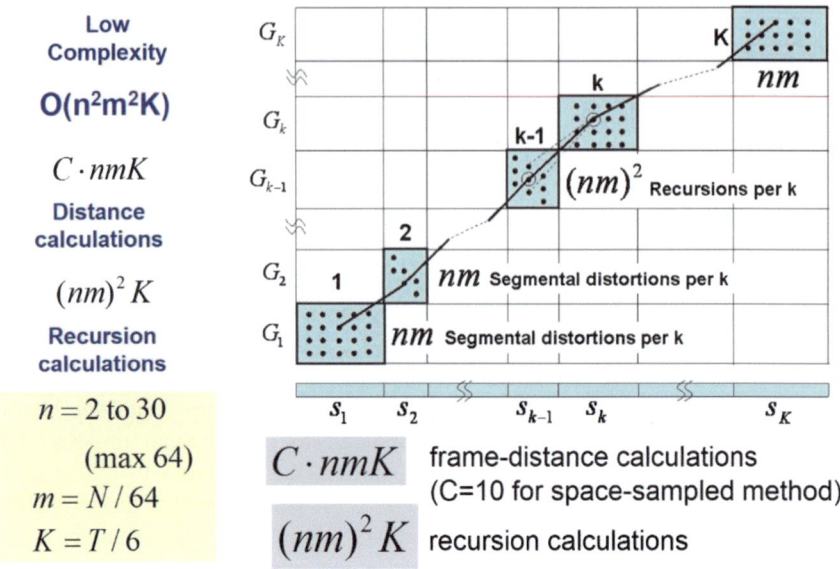

Fig. 5.6 Trellis search space and recursions of the proposed n-best lattice based unit-selection and associated complexity

Table 5.1 Relative complexities of 1-pass DP, Lee and Cox [LC02] and proposed n-best lattice based unit-selection algorithms

Algorithm	Distance calculations	Recursion calculations	Overall complexity	Equivalence
One-pass DP	MNT	$(3MN+N^2)T$	$O(N^2T)$	$O(N^2T)$
Lee and Cox [LC02]	CmK	m^2K	$O(m^2K)$	$O(N^2T/L^2M)$
n-best Lattice	CnmK	$(nm)^2 K$	$O(n^2m^2K)$	$O(n^2N^2T/L^2M)$

reduced by a factor of L^2M for the Lee-Cox algorithm and by L^2M/n^2 for the n-best lattice based algorithm.

Table 5.2 shows the different variables involved in the complexity calculations, and their typical values, so as the help realize actual numbers of these complexities. Such a typical numerical quantification of the overall complexity of the three algorithms is further shown in Table 5.3. Table 5.3 shows the overall complexity of the three algorithms as in the fourth column of Table 5.1, with the following parameters: $N = 2^{17}$ and $T = 50$ (1 s of input utterance), $L = 64$ ($m = 2^{11}$), $K = 10$ (10 segments/s also obtained as $\sim T/M$ for $M = 6$ as average), and $n = 10$ (this value of n is chosen considering the corresponding rate-distortion performance of the n-best lattice, which reaches close to that of the optimal 1-pass DP algorithm's rate distortion performance). It can be seen that while the 1-pass DP algorithm has a high complexity of $O(2^{40})$, the Lee and Cox [LC02] algorithm has a relatively very low complexity of $O(2^{26})$, which is many orders of magnitude less than the 1-pass

5.5 Proposed n-Best Lattice Search

Table 5.2 Variables in the complexity calculations

Number of units	N	2^{17}–2^{20}
Average number of frames/unit	M	4–8
Number of frames in i/p utterance (1 s)	T	50
Size of VLSQ clustered codebook	L	64
n-best lattice factor	n	2–30 (max 64)
Segment to frame factor (space sampling)	C	10

Table 5.3 Numerical order complexity of the three algorithms: 1-pass DP, Lee and Cox [LC02], and n-best lattice based unit-selection

High complexity of optimal one-pass DP unit selection	$O(2^{40})$
Low complexity of Lee and Cox unit selection	$O(2^{26})$
For 10-best (n = 10), low complexity of n-best unit selection	$O(2^{32})$

DP algorithm. The complexity of the n-best lattice based unit-selection is $O(2^{32})$, which is higher than Lee and Cox algorithm, but a factor of 2^6 less than the 1-pass DP algorithm.

The reduced complexity of the n-best lattice relative to the 1-pass DP algorithm and its relative marginal increase over the Lee and Cox algorithm can be visualized more comprehensively in the figures for complexity, Figs. 5.7 and 5.8 which show respectively the distance calculation complexity (column 2 in Table 5.1) and the recursion complexity (column 3 in Table 5.1, also treated as the overall complexity due to its quadratic dependence on the unit-database size N), as function of N, for the three different algorithms. The following can be noted from the figure:

1. In both the figures, the high complexity of the 1-pass DP algorithm (red-squared solid line) and particularly its rapid increase with N (linear with N for distance calculation in Fig. 5.7 and quadratic with N for recursion calculations in Fig. 5.8) can be noted.
2. In contrast, the relatively and dramatically low complexity of the Lee and Cox [LC02] algorithm (blue-squared solid line) and its significant independence on N can be noted in both these figures.
3. The complexity of the n-best lattice algorithm (dashed lines for various n) can be noted to lie in between these two lines, with a general trend of increasing complexity with n. While the distance calculations tend to seem to increase faster with n, the dominant recursion complexity (quadratic in N for all the three algorithms, but scaled down by a factor of L^2M for the Lee and Cox [LC02] and n-best lattice algorithms) is significantly low, particularly relative to the 1-pass DP algorithm's complexity, wherein the relative differences are more significant owing to the rapid increase in the 1-pass DP algorithm's un-factored quadratic complexity.

In summary, it seems reasonable to conclude from the results presented above (for n-best lattice based unit-selection for n up to 30), that the additional

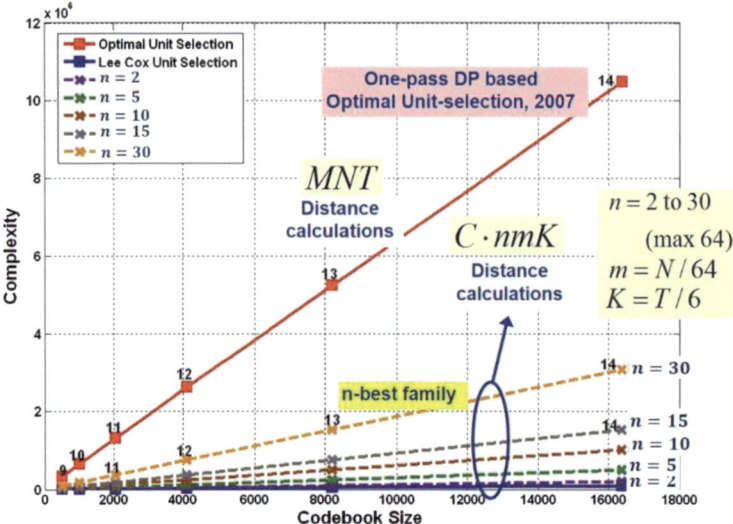

Fig. 5.7 Distance calculations complexity of 1-pass DP unit-selection, Lee and Cox [LC02] unit-selection and n-best lattice based unit-selection (for n = 2–30) algorithms

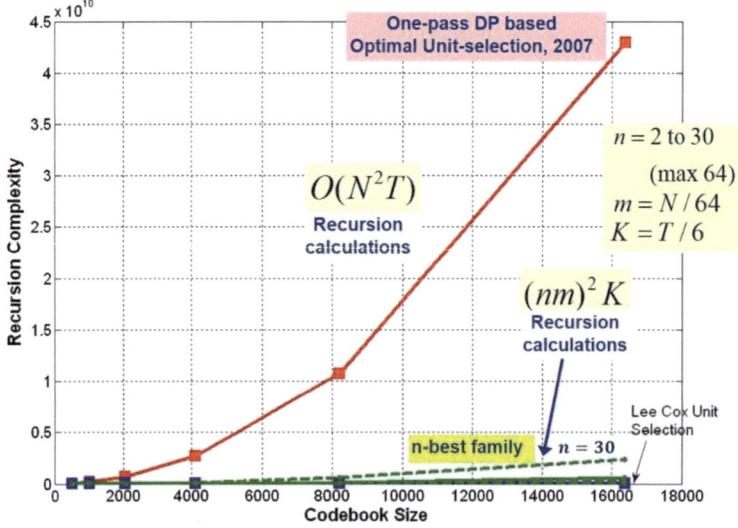

Fig. 5.8 Recursion complexity of 1-pass DP unit-selection, Lee and Cox [LC02] unit-selection and n-best lattice based unit-selection (for n = 2–30) algorithms

multiplicative cost of n to the distance calculation cost and n^2 to the recursion cost do not increase the computational complexity of the n-best lattice search significantly above that of the basic Lee and Cox unit-selection algorithm. This is satisfactory in the sense that the results do show a significant lowering of spectral distortion (or enhancement of the optimality) for reasonable n in the n-best lattice formulation without proportional increase in the complexity; or in other words, the rate-distortion performance of the n-best lattice search seems to reach those of the optimal 1-pass DP unit-selection algorithm, even while retaining the low complexity of the Lee and Cox segmental unit-selection framework.

5.6 Conclusions

We have proposed an n-best lattice based unit-selection algorithm that has near-optimal R-D performance as the fully optimal one-pass DP algorithm at a highly reduced complexity for practical ultra low bit-rate coding.

Chapter 6
No Residual Transmission: Joint Spectral-Residual Quantization

In this chapter, we present a unit-selection based segment quantization scheme which leads to the interesting possibility of not having to transmit any side-information about the residual at all [RH09], [R12]. This chapter elaborates on these two earlier work in a coherent manner, with several parts reproduced here with permission. We arrive at this scheme from the important observations (i) that unit-selection based segment quantization systems typically employ large unit-databases (as termed in concatenative speech synthesis) and, (ii) that by virtue of the largeness of the continuous codebook, it becomes possible to quantize an input segment by an unit in the unit database in such a way that the speech corresponding to the unit (after applying 'only' duration modification) is a close reconstruction of the input speech (of that input segment). This is realized in a LPC synthesis framework as follows: (i) apply duration modification on the unit quantizing the input segment to match the duration of the input segment, (ii) the speech corresponding to the (duration modified) unit is synthesized at the decoder using the LP frames of that unit and the corresponding residual in the unit-database without requiring any information about the input residual.

Figure 6.1 shows the generic block diagram of such a scheme, set in a LP analysis-synthesis framework. The encoder and decoder employ a large continuous codebook ('unit-database') as required for the unit-selection framework. The codebook is shown to be made of variable length segmental units, derived by either manual phonetic labeling (as in TIMIT type of databases) or by using the variable length segment quantization algorithm with a clustered codebook (such as discussed earlier in Chap. 3); the codebook is however, essentially a continuous codebook, typically derived from a long continuous speech (e.g. 6 h in the examples to be discussed later in this chapter), but segmented into variable length segments, as shown in this figure. With reference to such a codebook as used in Chap. 4, the main difference is, here the codebook is made of two components—the spectral codebook and the residual codebook—where each variable length segmental unit in the codebook consisting of a sequence of frame, has an associated sequence of

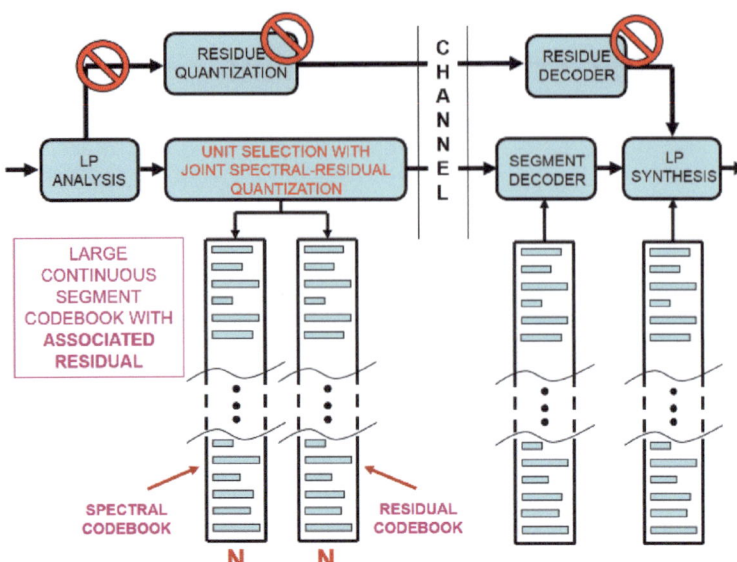

Fig. 6.1 Unit-selection framework with joint spectral-residual quantization employing a large continuous codebook made of spectral and residual components

residuals, one for each frame in the unit. As noted above, note how the residual quantization and transmission part is non-functional here.

In order to ensure that a selected unit in the unit-database can indeed approximate the input speech by using its own residual for synthesis, we propose a joint spectral-residual quantization scheme using various 'composite measures' for the unit-selection based segment quantization. These measures quantify how well both the spectra (LP envelope) and the residual (excitation component) of an unit (i.e., the frames in the unit) match those of the original input segment, thereby providing a match of the overall speech signal and ensuring that the synthesis at the decoder using the unit in totality (with its own residual) does indeed produce a speech signal 'close' to the input speech signal. By the proposed approach, we incur only the bit-rates for spectral and duration quantization and are able to limit the overall bit-rate to 250 bits/s for a continuous codebook of size 19 bits/unit (i.e., 524,288 units or about 6 h of speech).

A related work in this direction is the 'waveform segment vocoder' [RW85], where the templates selected for quantizing the input segment is represented at the decoder as a waveform (and not the spectral information), and the pitch, gain and duration of this 'waveform template' are modified to match those of the input segment. The modified segments are then concatenated to produce the output waveform. However, this method involved intricate pitch and duration modification in a RELP framework, where the waveform segment was decomposed into an LPC envelope and its corresponding residual and the residual is up sampled, low-pass filtered and down sampled by a pitch scaling factor with the resultant residual being both pitch-scale modified, and time-scale modified by an overlap-add procedure for linear time-warping. However, no overall operating bit-rate or spectral distortion was reported, but this work concluded that the initial results indicate that such an approach has potential for transmitting speech at rates below 300 bits/s.

In the approach proposed here, we are able to achieve a similar 'no-residue' transmission, but with far less complex processing on the units than the approach above, and show actual bit-rates of less than 300 bits/s (and associated spectral distortions) and consequently more accurate reconstruction of speech. This in a way corroborates the expectations of this much earlier work, but in a far more elegant, pleasing and computationally simpler framework set in the recent paradigm of unit-selection based quantization.

6.1 Joint Spectral-Residual Quantization in Lee-Cox Unit-Selection Framework

Figure 6.2 shows the proposed joint spectral-residual quantization (using composite distance measures) set in the unit-selection framework proposed earlier [LC02]. Note that the structure of the system is essentially same as in the basic algorithm described in Chap. 3 (Fig. 3.2), except for the definition of the 'composite' unit database, which is now shown to have a 'corresponding residual' component along with the main 'unit database' which has the spectral component. While the details of the steps in the unit-selection procedure of Lee and Cox [LC02] are as in the algorithm in Sect. 3.2, we give below the essential definitions again, in order to further develop the joint spectral-residual quantization formulation in a self-contained manner here.

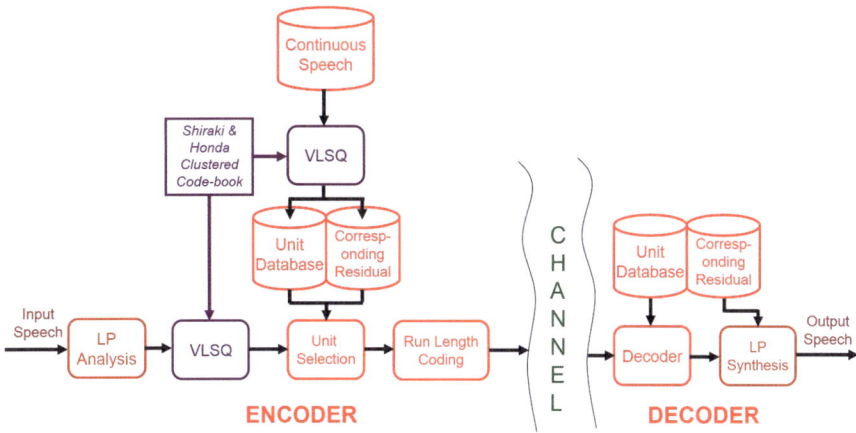

Fig. 6.2 Proposed 'no-residue transmission' unit-selection framework with joint spectral-residual quantization (Reused with permission from [RH09])

Composite Unit Database As defined before, the spectral component of the unit database is a long continuous speech with continuous sequences of LP parameter vectors. The unit-database is derived from this continuous speech by segmenting and quantizing (i.e., labeling) the continuous speech using a 'clustered' codebook $\mathcal{V} = \{v_1, v_2, \ldots, v_p, \ldots, v_P\}$ by variable-length segment quantization (VLSQ) [SH88]. This results in a unit-database which is a sequence of variable-length units $\mathcal{U} = (u_1, u_2, \ldots, u_N)$, where a unit u_n is of length l_n frames, given by $u_n = (u_n(1), u_n(2), \ldots, u_n(l_n))$; each unit has an index (or label) from the clustered codebook $\mathcal{V}$, i.e., the label of unit u_n is $\mathcal{L}(u_n) \in [1, \ldots, P]$. Typical value of P is 64. The residual component corresponding to each unit u_n is as follows: Each unit $u_n = (u_n(1), u_n(2), \ldots, u_n(l), \ldots, u_n(l_n))$ of l_n frames has a sequence of associated residual $r_n^u = (r_n^u(1), r_n^u(2), \ldots, r_n^u(l), \ldots, r_n^u(l_n))$ where the frame with the LP vector $u_n(l)$ has its own residual $r_n^u(l)$. As we will see in Sect. 6.1.1, this enables synthesis at the decoder using the units in the unit database in a self-contained manner, without requiring the input residual to be modeled, quantized and transmitted.

Input Speech Segmentation and Unit Grouping As in the unit-selection procedure of Lee and Cox'02 [LC02] described in detail in Sect. 3.2, the input speech utterance $O = (o_1, o_2, \ldots, o_t, \ldots, o_T)$ is segmented and labeled by the clustered codebook $\mathcal{V}$ to yield a segmentation of K segments given by $S = (s_1, s_2, \ldots, s_{k-1}, s_k \ldots, s_K)$ with corresponding segment lengths $(L_1, L_2, \ldots, L_{k-1}, L_k \ldots, L_K)$. Each segment s_k is associated with a label from the VLSQ codebook $\mathcal{V}$, denoted by $\mathcal{L}(s_k)$. Following the input segmentation, a unit group $\mathcal{G}_k$ is computed corresponding to each $s_k, k = 1, \ldots, K$ where $\mathcal{G}_k$ is a collection of all units in the unit database $\mathcal{U}$ such that the VLSQ label of all these units are the same as that of s_k, i.e., $\mathcal{G}_k = \{u_n : \mathcal{L}(u_n) = \mathcal{L}(s_k)\}$, where the units in group $\mathcal{G}_k$ are the potential candidate units for quantizing segment s_k.

Unit-Selection Formulation With the definition of $S = (s_1, s_2, \ldots, s_{k-1}, s_k \ldots, s_K)$ and $\mathcal{G}_k, k = 1, \ldots, K$ as above, it is now required to determine the optimal sequence of unit indices $Q^* = (q_1^*, q_2^*, \ldots, q_{k-1}^*, q_k^*, \ldots q_K^*)$ that minimize an overall decoding distortion (quantization error) when the segment sequence $S = (s_1, s_2, \ldots, s_{k-1}, s_k \ldots, s_K)$ is quantized by the corresponding unit sequence $\left(u_{q_k^*}, k = 1, \ldots, K\right)$.

The overall distortion (quantization error) in quantizing S by any Q is given by

$$D^* = \min_Q \left[\alpha \sum_{k=1}^{K} D_u(s_k, u_{q_k}) + (1-\alpha) \sum_{k=2}^{K} D_c(q_{k-1}, q_k) \right] \quad (6.1)$$

with the corresponding optimal unit sequence $Q^* = (q_1^*, q_2^*, \ldots, q_{k-1}^*, q_k^*, \ldots q_K^*)$ that quantizes $s_k, k = 1, \ldots, K$ being

$$Q^* = \arg\min_Q \left[\alpha \sum_{k=1}^{K} D_u(s_k, u_{q_k}) + (1-\alpha) \sum_{k=2}^{K} D_c(q_{k-1}, q_k) \right] \quad (6.2)$$

Here, $D_u(s_k, u_{q_k})$ is the unit-cost (or distortion) in quantizing segment s_k using unit u_{q_k} which is restricted to belong to $\mathcal{G}_k$. $D_c(q_{k-1}, q_k)$ is the concatenation-cost (or distortion) when unit $u_{q_{k-1}}$ is followed by unit u_{q_k}, but with the restriction that $u_{q_{k-1}}, u_{q_k}$ belong respectively to groups $\mathcal{G}_{k-1}$ and $\mathcal{G}_k$, i.e., $u_{q_{k-1}} \in \mathcal{G}_{k-1}$ and $u_{q_k} \in \mathcal{G}_k$. Further, the actual units $u_{q_{k-1}}, u_{q_k}$ selected from these groups are subject to the concatenation constraint

$$D_c(q_{k-1}, q_k) = \beta_{k-1,k} \cdot d(u_{q_{k-1}}(l_{q_{k-1}}), u_{q_k}(1)) \qquad (6.3)$$

where, $d(u_{q_{k-1}}(l_{q_{k-1}}), u_{q_k}(1))$ is the Euclidean distance between the last frame of unit $u_{q_{k-1}}$ and the first frame of unit u_{q_k}. $\beta_{k-1,k} = 0$ if $q_k = q_{k-1} + 1$ (i.e., $u_{q_{k-1}}$ and u_{q_k} are consecutive in the unit database) and $\beta_{k-1,k} = 1$ otherwise. This favors quantizing two consecutive segments (s_{k-1}, s_k) with two units which are consecutive in the codebook; run-length coding further exploits such 'contiguous' unit sequences to achieve lowered bit-rates.

Duration Modification In addition to the basic formulation as above, the joint spectral-quantization to be discussed in the next section stems from the definition of 'duration modification' that is to be done at the decoder to make the unit u_{q_k} (of length l_{q_k}) to match the duration of the input segment s_k (of length L_k) that it quantizes, as well as in the computation of the unit cost $D_u(s_k, u_{q_k})$ at the encoder above. This involves time warping of the unit u_{q_k} (of length l_{q_k}) to match the duration of the input segment s_k (of length L_k). While Lee and Cox [LC02] employ a bi-linear interpolation to carry out this duration modification, we do this here by 'space-sampling' - a resampling of unit u_{q_k} to yield L_k frames so as to match the duration of s_k, which is as follows: The total length of unit u_{q_k} is computed (using a Euclidean norm on the LP vectors (say, LARs) of u_{q_k}; then the spectral trajectory of the unit u_{q_k} is resampled at L_k equispaced points in the LAR space. Thus, if the segment s_k is given by the sequence of frames $s_k(1), s_k(2), \ldots, s_k(l), \ldots, s_k(L_k)$ and the unit u_{q_k} is given by the sequence of frames $u_{q_k}(1), u_{q_k}(2), \ldots, u_{q_k}(l), \ldots u_{q_k}(l_{q_k})$, the resampled version of the unit u_{q_k} is given as $u'_{q_k} = u'_{q_k}(1), u'_{q_k}(2), \ldots, u'_{q_k}(l), \ldots u'_{q_k}(L_k)$, wherein the re-sampled unit u'_{q_k} is now matched in duration with the input segment s_k. The unit cost $D_u(s_k, u_{q_k})$ is then defined as

$$D_u(s_k, u_{q_k}) = \sum_{l=1}^{L_k} d(s_k(l), u'_{q_k}(l)) \qquad (6.4)$$

6.1.1 Joint Spectral-Residual Quantization

The essential principle of joint spectral-residual quantization is based on the fact that the output speech is synthesized using the unit sequence $(u'_{q_1^*}, u'_{q_2^*}, \ldots, u'_{q_k^*}, \ldots, u'_{q_K^*})$ where each frame's LP filter is driven by its own corresponding residual without having any side information of the residual transmitted to the decoder. The optimization involved in determining Q^* therefore has to solve for Q^* by using an appropriate 'composite distance measure' in defining $D_u(s_k, u_{q_k})$, so that the resulting minimized distortion reflects the best match between the input segments $s_k, k = 1, \ldots, K$ with the corresponding units $u_{q_k}, k = 1, \ldots, K$. Each resultant space sampled unit $u'_{q_k}, k = 1, \ldots, K$ can then be used for LP synthesis with its own residual in a self-contained manner without requiring any side-information about the residual (of the frames of the input segments s_k) to be transmitted to the decoder.

The main modification to the above framework for realizing a joint spectral-residual quantization lies in defining $D_u(s_k, u_{q_k})$ appropriately to take into account the matching of both the spectral properties (as defined by the LP parameters) and the residual properties by means of 'composite distance measures' so as to arrive at an overall match between the speech signal corresponding to segment s_k and unit u_{q_k}. Such composite distance measures consider the residual information along with the spectral information in different ways.

For defining the composite distance measures, let the segment $s_k = (s_k(1), s_k(2), \ldots, s_k(l), \ldots, s_k(L_k))$ be associated with its residual frames $r^i_k = (r^i_k(1), r^i_k(2), \ldots, r^i_k(l), \ldots, r^i_k(L_k))$ where the superscript i stands for input segment. Likewise, let the resampled unit $u'_{q_k} = \left(u'_{q_k}(1), u'_{q_k}(2), \ldots, u'_{q_k}(l), \ldots u'_{q_k}(L_k)\right)$ have its associated residual frames as $r^u_{q_k} = \left(r^u_{q_k}(1), r^u_{q_k}(2), \ldots, r^u_{q_k}(l), \ldots, r^u_{q_k}(L_k)\right)$ where the superscript u stands for unit. Note that one way to obtain the residual frame $r^u_{q_k}(l)$ (i.e., the lth frame in the sequence of residuals $r^u_{q_k}$) corresponding to the resampled frame $u'_{q_k}(l)$ (i.e., the lth frame in the resampled unit u'_{q_k}) is as follows: $u'_{q_k}(l)$ is made to inherit its residual $r^u_{q_k}(l)$ from the original frame $u_{q_k}(l')$ that is closest to $u'_{q_k}(l)$ in terms of frame-to-frame spectral distortion. Figure 6.3 illustrates this with a schematic—the duration modified unit u'_{q_k} with L_k frames is shown matched with the input segment s_k with L_k frames. Each frame in s_k and u'_{q_k} is shown to have two components—the spectral component (vertical rectangle) and the residual component (circle).

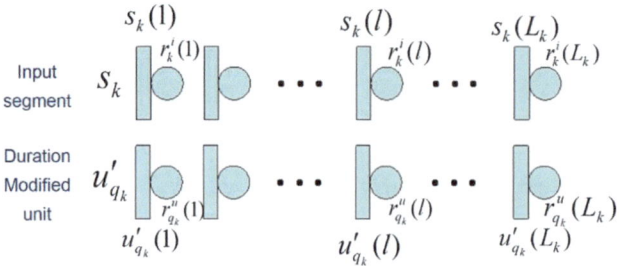

Fig. 6.3 Schematic of the duration modified unit matching with the original input segment. Each frame is shown to have two components—rectangle (spectral component) and circle (residual component)

6.1 Joint Spectral-Residual Quantization in Lee-Cox Unit-Selection Framework

We now consider five definitions of the unit cost $D_u(s_k, u_{q_k})$ in Eq. (6.4) based on the underlying distance measures used in matching the corresponding frames in input segment s_k and unit u_{q_k}. Of these, the first is the standard LAR based Euclidean distance and is not really a 'composite' measure in the sense that it measures 'only' the spectral quantization error and is used here mainly as a baseline to represent a conventional system which does not concern matching the residual also. The other four definitions are 'composite measures' in the sense they combine spectral and residual information in different ways to match the overall spectra (speech).

1. **LAR only (LAR):** $s_k(l)$ and $u'_{q_k}(l)$ are LAR vectors (of dim d) given by $s_k(l) = (s_k(l,j), j = 1, \ldots, d)$ and $u'_{q_k}(l) = \left(u'_{q_k}(l,j), j = 1, \ldots, d\right)$ and $d\left(s_k(l), u'_{q_k}(l)\right)$ is the Euclidean distance between them, given by

$$\sum_{j=1}^{d} \left| s_k(l,j) - u'_{q_k}(l,j) \right|^2$$

2. **High dimension MFCC (MFCC):** $s_k(l)$ and $u'_{q_k}(l)$ are MFCC vectors of high dimension (80) derived by using 80 triangular filters and all the 80 coefficients after DCT. $d\left(s_k(l), u'_{q_k}(l)\right)$ is the Euclidean distance between them. The high dimension MFCC is intended to represent both the spectral and residual information by virtue of its larger number of finely spaced filters. A dimension of 80 was found optimal (among values 40, 80, 120, 160) for best performance.

3. **LAR and Pitch (LAR + P):** This combines the LAR based distance metric with pitch differences between corresponding frames of input segment s_k and unit u'_{q_k} given by

$$d\left(s_k(l), u'_{q_k}(l)\right) = \gamma \sum_{j=1}^{d} \left| s_k(l,j) - u'_{q_k}(l,j) \right|^2 + (1-\gamma) \left| P^i_k(l) - P^u_{q_k}(l) \right|^2 \quad (6.5)$$

where $P^i_k(l)$ is the pitch value of the lth frame of input segment s_k and $P^u_{q_k}(l)$ is the pitch value of the lth frame of the re-sampled unit u'_{q_k}. The pitch values are zero for unvoiced segments/units. $\gamma = 0.6$ was used for best performance.

4. **Log magnitude power spectral distortion (PSD):** Here, $s_k(l)$ and $u'_{q_k}(l)$ are the log magnitude power spectra (256 point spectra) of the lth frame of the input segment s_k and unit u'_{q_k}. Such an attempt to match the power spectra takes into account both the LP spectra and the excitation spectra together, thereby providing a direct match between the input speech and the unit speech segments. $d\left(s_k(l), u'_{q_k}(l)\right)$ is given by

$$d\left(s_k(l), u'_{q_k}(l)\right) = \sum_{j=1}^{256} \left| s_k(l,j) - u'_{q_k}(l,j) \right|^2 \quad (6.6)$$

5. **Cross-correlation (CC):** Here we define $d\left(s_k(l), u'_{q_k}(l)\right)$ as a weighted sum of the Euclidean distance $d'\left(s_k(l), u'_{q_k}(l)\right)$ between the LAR vectors ($s_k(l)$ and $u'_{q_k}(l)$) and one minus the best cross-correlation value between their corresponding residual $r_k^i(l)$ and $r_{q_k}^u(l)$. This is given by

$$d\left(s_k(l), u'_{q_k}(l)\right) = \gamma\, d'\left(s_k(l), u'_{q_k}(l)\right) + (1-\gamma)\left(1 - CC\left(r_k^i(l), r_{q_k}^u(l)\right)\right) \quad (6.7)$$

where, $CC\left(r_k^i(l), r_{q_k}^u(l)\right)$ is given by [SS92]

$$CC\left(r_k^i(l), r_{q_k}^u(l)\right) = \max_m \rho(m) \quad (6.8)$$

$$\rho(m) = \frac{\langle a(i)b(i-m)\rangle - \langle a(i)\rangle\,\langle b(i-m)\rangle}{\left(\langle a^2(i)\rangle - \langle a(i)\rangle^2\right)^{\frac{1}{2}}\left(\langle b^2(i)\rangle - \langle b(i)\rangle^2\right)^{\frac{1}{2}}} \quad (6.9)$$

where $a(\cdot)$ and $b(\cdot)$ are used to represent respectively $r_k^i(l)$ and $r_{q_k}^u(l)$ (for notational convenience), m is the positive window shift of the unit residual $r_{q_k}^u(l)$ in computing the cross-correlation, $\langle\,\rangle$ denotes the time average of the enclosed expression for i ranging over the window length. Note that the index i used in Eq. (6.9) is not to be confused with the super-script i representing 'input' in the input residual $r_k^i(l)$ in Eq. (6.8). $\rho(m)$ is the normalized cross-correlation defined and used earlier in [SS92] between the synthetic glottal volume velocity (obtained from a glottal model) and that obtained from inverse filtering the original speech as part of a set of cost functions minimized in estimating articulatory parameters from a given speech signal. This was perhaps a rare instance of a measure that matches residual in time domain and suits the composite measure required here. $\gamma = 0.5$ was used for best performance.

By using these 'composite measures', the attempt is to match the spectra (LP envelope) and the residual jointly in order to give a measure of the overall match of the input segment s_k and unit u'_{q_k}. Selection of a unit u_{q_k} for quantizing segment s_k through a good match under the composite measure then indicates that the use of the residual $r_{q_k}^u(l)$ in LP synthesis with filter parameters (prediction coefficients) derived from $u'_{q_k}(l)$ can yield a speech signal frame very close to the speech signal corresponding to the frame $s_k(l)$ in the input segment s_k and consequently in being able to synthesize an output speech segment that closely resembles $s_k, \forall k = 1, \ldots, K$.

6.1.2 Experiments and Results

The proposed unit-selection with joint spectral-residual quantization using the five different composite measures was evaluated on a unit-database derived from an audio book 'The Red-Limit' (http://www.harpercollins.com/books/9780061632112/Red_Limit_The/index.aspx) which is a single speaker database comprising of passages in the domain of astronomy spoken in English by a male speaker with an American accent. The unit-database was divided into two parts, Train and Test, with about 6 h of speech in Training and 1 h of speech in Test. We used unit-databases from Train with sizes ranging from 9 bits/unit (512 units) to 19 bits/unit (524,288 units or the full 6 h of speech). We report results on one test sentence ('By which time the sun has moved across the face of our galaxy') from Test of about 5 s duration (same speaker as Train but outside all the unit-databases from Train).

We present results of the proposed unit-selection based segment quantization in terms of quantization accuracy using rate-distortion curves between conventional average spectral distortion (as defined in Sect. 2.8) and the overall bit-rate with run-length coding. This is shown in Fig. 6.4 for the five composite measures.

The conventional average spectral distortion (SD) is measured between the original sequence of linear-prediction vectors and the sequence obtained after segment quantization and duration modification at the decoder. The average spectral distortion is the average of the single frame spectral distortion over the number of frames in the input speech; the single frame spectral distortion is the squared difference between the log of the linear-prediction power spectra of the original frame and the quantized frame, averaged over frequency. This essentially measures

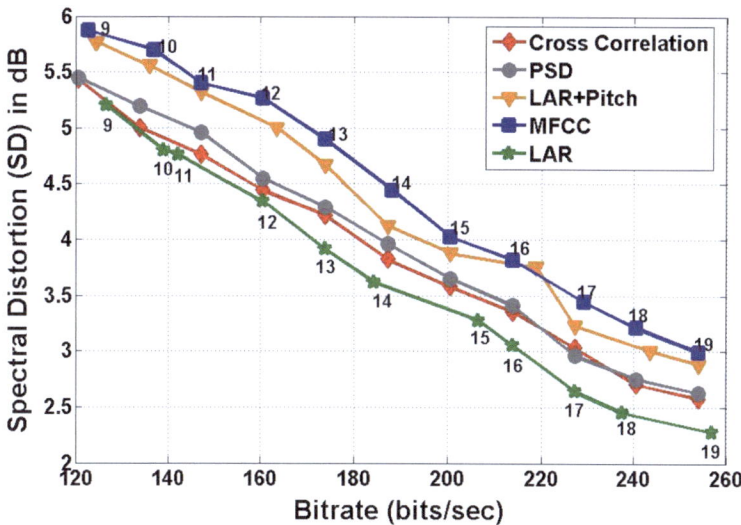

Fig. 6.4 Rate-distortion plot (SD vs. overall bit-rate) (Reused with permission from [RH09])

how well the spectral information is quantized under the joint spectral-residual quantization scheme.

The following can be noted:

(a) LAR performs the best since it matches only the spectra; the other four composite measures perform with slightly higher spectral distortion since they emphasize matching both the spectra and residual and in the process compromise on the spectral match alone.
(b) Of the other composite measures, CC and PSD perform the next best indicating they do manage to achieve a good spectral match too. LAR + P and MFCC have higher distortions.
(c) The spectral distortion shows a prominent reduction with bit-rate and reaches as low as 2.5 dB (for LAR and CC) for effective overall bit-rates of 250 bits/s. (As a reference, 1 dB spectral distortion corresponds to the standard transparent quality [PK95]).

Since our objective is to compare the 5 different composite measures used under the 'no-residue transmission' framework, it is meaningful to show the 'overall' performance in terms of the average log magnitude power spectral distortion (PSD) between the input speech frames and output speech frames vs. overall bit-rate. This is shown in Fig. 6.5 for the five composite measures. For PSD, the single frame power spectral distortion is the squared difference between the log of the magnitude power spectra of the original speech frame and the output synthesized speech frame.

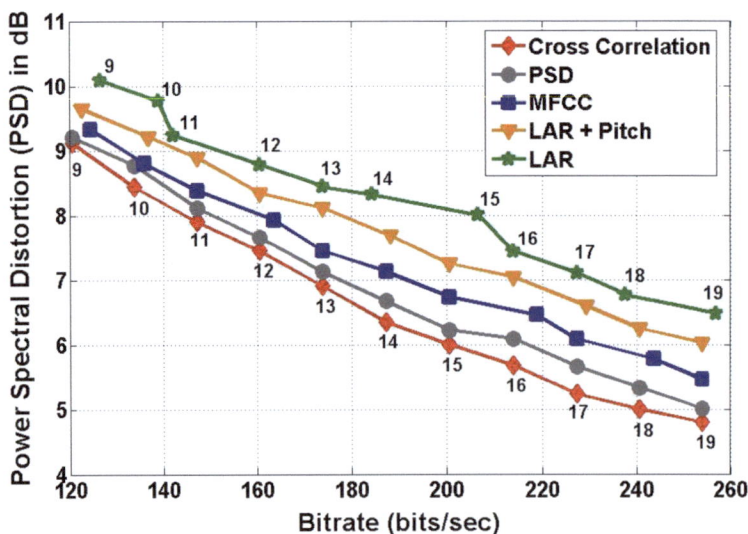

Fig. 6.5 Rate-distortion plot (PSD vs. overall bit-rate) (Reused with permission from [RH09])

6.1 Joint Spectral-Residual Quantization in Lee-Cox Unit-Selection Framework

The following can be observed from this figure:

(a) The PSD values are higher than SD values since the overall PSD includes the residual spectra mismatch (over and above LP envelope spectral differences),
(b) CC performs the best indicating that it achieves the best match of spectra and residual combined,
(c) This is followed by PSD and then by MFCC indicating their ability to match the overall spectra (including residual) quite adequately,
(d) LAR performs the poorest indicating it is unable to select units with the residual also matching; LAR + P performs better indicating the addition of pitch is able to do a better composite match in selecting units with the right residual by virtue of using just a single prosody parameter, pitch.

The above observations, when interpreted in a combined way (SD and PSD), corroborates with the actual quality of the synthesized speech by listening. While formal listening tests are underway, we found that CC performs the best in terms of overall quality of speech which is intelligible, natural and retains speaker-identity and prosody (gain and intonation) and is closest to the input speech; this is borne out by the fact the CC has the lowest PSD and second best SD (very close to the SD that LAR alone can achieve). This shows the feasibility of realizing practical ultra low bit-rate speech coding at 250 bits/s under the joint spectral-residual quantization scheme in the unit-selection framework using unit-database of size 19 bits/unit (or about 6 h of data). Since the SD continues to decrease at a steady rate, it can be expected that use of larger unit-databases (perhaps 10 h or more as in concatenative TTS) can yield spectral distortions as low as 2 dB and less.

This is also borne out by the plots of speech waveform (Fig. 6.6) and residual (Fig. 6.7) for the five composite measures. The selected speech is the diphthong /ay/

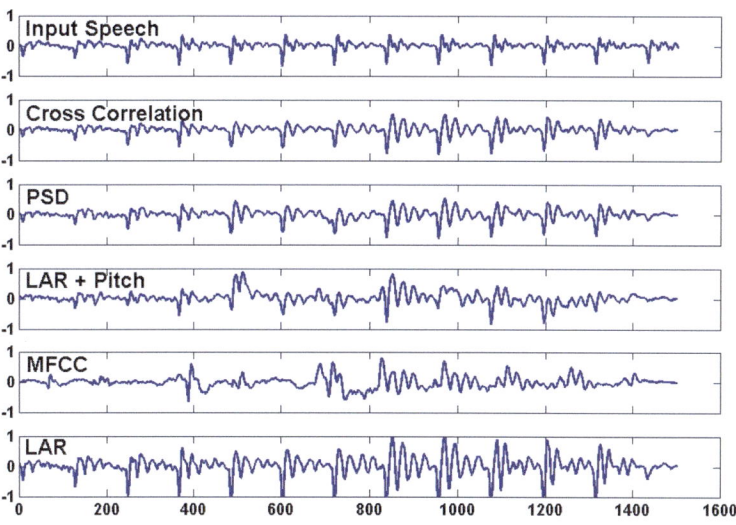

Fig. 6.6 Input speech and output speech waveforms (Reused with permission from [RH09])

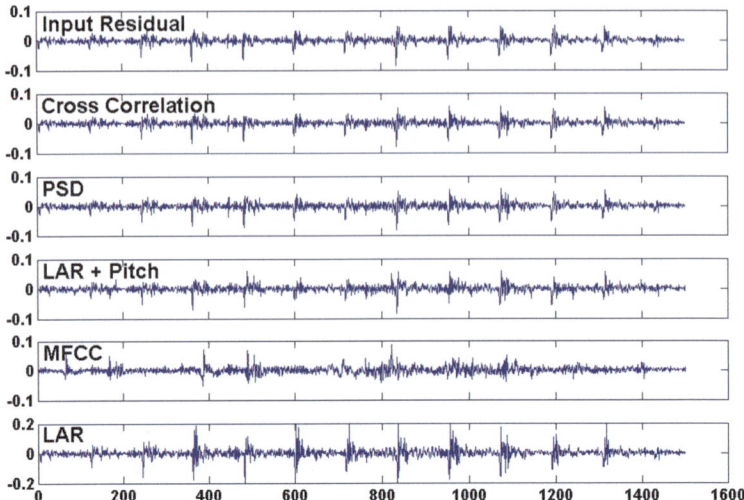

Fig. 6.7 Input residual and unit-database residuals used in synthesis (Reused with permission from [RH09])

from the word 'time'. Clearly, CC performs closest to the original input speech, with a remarkable match of the residue with differences in the region of fourth and fifth pitch period between 1,000 and 1,500. The non-exact match of the speech waveform (given the more or less exact match of the residual) is due to the spectral quantization errors. PSD is close to CC; LAR + P and MFCC perform poorly and LAR has some match due to the inherent coupling of the residual to the LP parameters (but with a gross mismatch in gain).

6.2 Joint Spectral-Residual Quantization in an Optimal Unit-Selection Framework

Recognizing the sub-optimality of the unit-selection framework of [LC02] used in our joint spectral-residual quantization [RH09] (discussed in Sect. 6.1), we present in this section the algorithm proposed by us in [R12], which uses joint spectral-residual quantization in our earlier optimal one-pass DP framework of [RH06, RH07], thereby leading to better rate-distortion performance of the overall 'no-residual-transmission' system than the results we obtained in [RH09]. Within the optimal one-pass DP based unit-selection segment quantization presented here, the 'composite distance measures' (underlying the 'joint spectral-residual quantization') are defined within the local distance used in the recursions of the one-pass DP unit-selection procedure. Here, these measures quantify how well both the spectra (LP envelope) and the residual (excitation component) of a frame of a unit

6.2 Joint Spectral-Residual Quantization in an Optimal Unit-Selection Framework 137

(in the unit database) match those of a frame of the original input speech, thereby providing a match of the overall speech signal and ensuring that the synthesis at the decoder using the unit in totality (with its own residual) does indeed produce a speech signal 'close' to the input speech signal.

6.2.1 Joint Spectral-Residual Quantization in 1-Pass DP Framework

The basic modifications to the 1-pass DP framework to carry out joint spectral-residual quantization are comparatively different than in the Lee and Cox sub-optimal framework described in Sect. 6.1. In order to further develop the basic formulation of the joint spectral-residual quantization in the 1-pass DP framework, but without repeating the basic formulation of the 1-pass DP unit-selection framework, we refer back to Sect. 4.1.1. All notations in the following treatment are consistent with the equations in Sect. 4.1.1.

Figure 6.8 shows the basic structure of the unit-selection based quantizer using joint-spectral residual quantization set in the 1-pass DP framework. Note the use of the composite database with the spectral component and the associated residual component. Also note that absence of the pre-quantization of the input speech, as is done in the Lee-Cox segmental unit-selection framework in Fig. 6.2. The clustered codebook used for variable length segment quantization (VLSQ), applied on the continuous speech is to render the continuous codebook into a variable-length 'unit-database'.

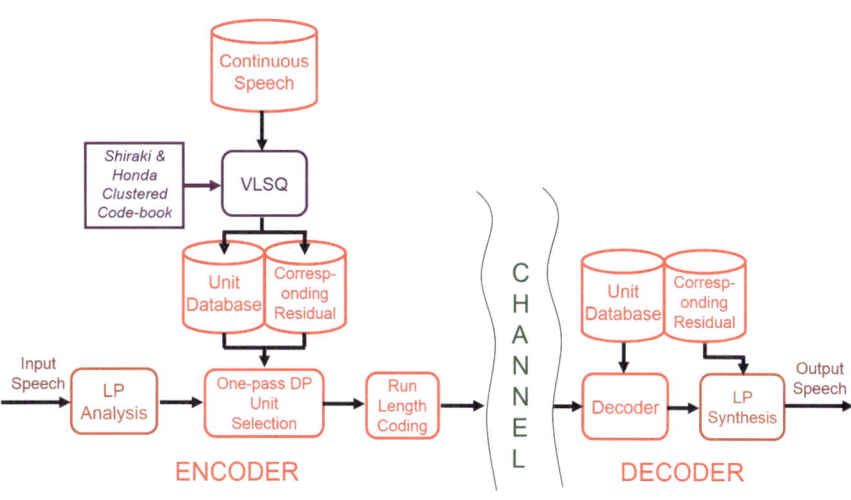

Fig. 6.8 Joint spectral-residual quantization set in the optimal unit-selection framework based on 1-pass DP algorithm (Reused with permission from [R12])

The main modification to the 1-pass DP framework (referred above, as in Sect. 4.1.1) for realizing a joint spectral-residual quantization lies in defining $D_u(s_k, u_{q_k})$ in Eq. (4.1) appropriately to take into account the matching of both the spectral properties (as defined by the LP parameters) and the residual properties by means of 'composite distance measures' so as to arrive at an overall match between the input utterance and the synthesis units (in the unit database). This in turn is realized by defining the local distance $d(i,j,n)$ in Eq. (4.3) and $d(i,1,n)$ in Eq. (4.4) as a 'composite distance measure' in such a way as to take into account the spectral mismatch and the residual mismatch between frame 'i' of input utterance and frame 'j' of unit 'n' in the unit-database.

In order to define such a composite distance measure $d(i,j,n)$, let frame 'i' in the input utterance be associated with spectral information S_i (say, a LAR vector of dim d) and residual information r_i. Likewise, let frame 'j' of unit n be associated with spectral information S_j^n and residual information r_j^n.

We had earlier proposed five different 'composite distance measures' [RH09], namely, (1) LAR only, (2) High dimension MFCC (MFCC), (3) LAR and Pitch (LAR+P), (4) Log magnitude power spectral distortion (PSD) and (5) Cross-correlation (CC). Of these, we had shown in Sect. 6.1 ([RH09]) that CC performed the best among the various composite distance measures; hence, we consider here only two of these measures, namely, (1) LAR only (LAR) and (2) Cross-correlation (CC) for defining $d(i,j,n)$ as a 'composite distance measure' in Eqs. (4.3) and (4.4). Of these, the first is the standard LAR based Euclidean distance and is not really a 'composite' measure in the sense that it measures 'only' the spectral quantization error and is used here mainly as a baseline to represent a conventional system which does not concern matching the residual also. CC is a 'composite measure' in the sense it combines spectral and residual information to match the overall spectra (speech).

1. **LAR only (LAR):** S_i and S_j^n are LAR vectors (of dimension d); i.e., $S_i = S_i(l)$, $l=1,\ldots,d$ and $S_j^n = S_j^n(l)$, $l=1,\ldots,d$. $d(i,j,n)$ is the Euclidean distance between them, given by,

$$d(i,j,n) = \sum_{l=1}^{d} \left| S_i(l) - S_j^n(l) \right|^2 \tag{6.10}$$

2. **Cross-correlation (CC):** Here we define $d(i,j,n)$ as a weighted sum of the Euclidean distance between the LAR vectors (S_i and S_j^n) and one minus the best cross-correlation value between their corresponding residual r_i and r_j^n. This is given by, ($\gamma = 0.5$ was used for best performance),

6.2 Joint Spectral-Residual Quantization in an Optimal Unit-Selection Framework

$$d(i,j,n) = \gamma \sum_{l=1}^{d} \left| S_i(l) - S_j^n(l) \right|^2 + (1-\gamma)\left(1 - CC\left(r_i, r_j^n\right)\right) \quad (6.11)$$

where $CC(r_i, r_j^n)$ is given by [SS92]

$$CC\left(r_i, r_j^n\right) = \max_m \rho(m) \quad (6.12)$$

$$\rho(m) = \frac{\langle a(i)b(i-m)\rangle - \langle a(i)\rangle \langle b(i-m)\rangle}{\left(\langle a^2(i)\rangle - \langle a(i)\rangle^2\right)^{\frac{1}{2}} \left(\langle b^2(i)\rangle - \langle b(i)\rangle^2\right)^{\frac{1}{2}}} \quad (6.13)$$

where $a(\cdot)$ and $b(\cdot)$ are used to represent respectively r_i and r_j^n (for notational convenience), m is the positive window shift of the unit residual r_j^n in computing the cross-correlation, $\langle \rangle$ denotes the time average of the enclosed expression for i ranging over the window length. Note that the index i used in Eq. (6.13) is not to be confused with the index i in $d(i,j,n)$ in Eq. (6.11) and sub-script i in Eqs. (6.11) and (6.12) representing the ith frame of input utterance. $\rho(m)$ is the normalized cross-correlation defined and used earlier in [SS92] between the synthetic glottal volume velocity (obtained from a glottal model) and that obtained from inverse filtering the original speech as part of a set of cost functions minimized in estimating articulatory parameters from a given speech signal. This was perhaps a rare instance of a measure that matches residual in time domain and suits the composite measure required here. $\gamma = 0.5$ was used for best performance.

By using these 'composite distance measures', the attempt is to match the spectra (LP envelope) and the residual jointly in order to give a measure of the overall match of the 'speech signal' of the input frame i and the unit frame j. Subsequent to the optimal decoding (segment quantization) by Eq. (4.1) and Eq. (4.5), selection of a unit $u_{q_k^*}$ for quantizing segment s_k through a good match under the composite measure then indicates that the use of the residual $r_j^{q_k^*}$ in LP synthesis with filter parameters (prediction coefficients) of frame j of unit $u_{q_k^*}$ can yield a speech signal frame very close to the speech signal corresponding to the frame i of segment s_k in the input utterance; this would hold good for every frame j of unit $u_{q_k^*}$ and further for all units $u_{q_k^*}, k = 1, \ldots, K$ thereby ensuring synthesis of output speech segments that closely resemble the corresponding input speech segments $s_k \forall k = 1, \ldots, K$.

The decoder performs LP synthesis using units $u_{q_k^*}, k = 1, \ldots, K$ (with their own associated residuals) after performing duration modification of unit $u_{q_k^*}$ to the length of input segment s_k as follows: The within-unit recursion Eq. (4.3) allows for reaching the trellis grid point corresponding to (i,j)], say $Q(i,j)$, (for any n) from the candidate grid points $P_1(i-1,j)$ or $P_2(i-1,j-1)$ or $P_3(i-1,j-2)$. This in turn implies that the optimal warping (part of the overall optimal path) between an optimal unit $u_{q_k^*}$ and an input segment s_k, ($\forall\, k = 1, \ldots, K$ making up the full decoding) can be specified as incremental steps of the kind $P(i-1, j+\Delta_i) \to Q(i,j)$, where $\Delta_i \in \{0, 1, 2\}$. This allows duration modification of unit $u_{q_k^*}$ (of length $l_{q_k^*}$) to segment s_k (of length L_k) along the optimal warping path by appropriately inserting/repeating/deleting frames of $u_{q_k^*}$ according to Δ_i to yield the quantized version of each frame i in the input segment. This duration modification results in a quantized version of input segment s_k using the frames of unit $u_{q_k^*}$ exactly as determined by the optimal one-pass DP based unit-selection at the encoder. This requires transmission of Δ_i using $\log_2 |\Delta_i|$ or 2 bits per frame i, resulting in an additional bit-rate of 100 bits/s (for a frame rate of 50 frames/s) over the overall bit-rate of the coder in Sect. 6.1 [RH09], which used space-sampling based duration modification (which is another source of sub-optimality). Figures 6.2 and 6.3 reflect this additional 100 bits/s in the optimal unit-selection algorithm proposed here.

6.2.2 Experiments and Results

We evaluate the proposed optimal unit-selection based on one-pass DP with joint spectral-residual quantization (OPT-1P-DP) using the 2 composite measures (LAR and CC) on a unit-database derived from an audio book and compare it with [RH09] (SUBOPT-LeeCox) set in the suboptimal framework [LC02] (Sect. 6.1). The audio book used ('The Red-Limit' (http://www.harpercollins.com/books/9780061632112/Red_Limit_The/index.aspx) is a single speaker database comprising of passages in the domain of astronomy spoken in English by a male speaker with an American accent. The unit-database was divided into 2 parts, Train and Test, with about 6 h of speech in Training and 1 h of speech in Test. We used unit-databases from Train with sizes ranging from 9 bits/unit (512 units) to 19 bits/unit (524,288 units or the full 6 h of speech). We report results on one test sentence ('By which time the sun has moved across the face of our galaxy') from Test of about 5 s duration (same speaker as Train but outside all the unit-databases from Train).

We present results comparing the proposed optimal one-pass DP based algorithm (OPT-1P-DP) and the earlier sub-optimal version (SUBOPT-LeeCox) using rate-distortion (R-D) curves for the two composite measures LAR and CC. Figure 6.9 shows the conventional average spectral distortion (SD) vs overall bit-rate with run-length coding for OPT-1P-DP (solid lines) and SUBOPT-LeeCox (dashed lines).

The conventional average spectral distortion (SD) (as defined in Sect. 2.8), is measured between the original sequence of linear-prediction vectors and the sequence obtained after segment quantization and duration modification at the decoder. The average spectral distortion is the average of the single frame spectral distortion over the number of frames in the input speech; the single frame spectral distortion is the squared difference between the log of the linear-prediction power spectra of the original frame and the quantized frame, averaged over frequency.

6.2 Joint Spectral-Residual Quantization in an Optimal Unit-Selection Framework

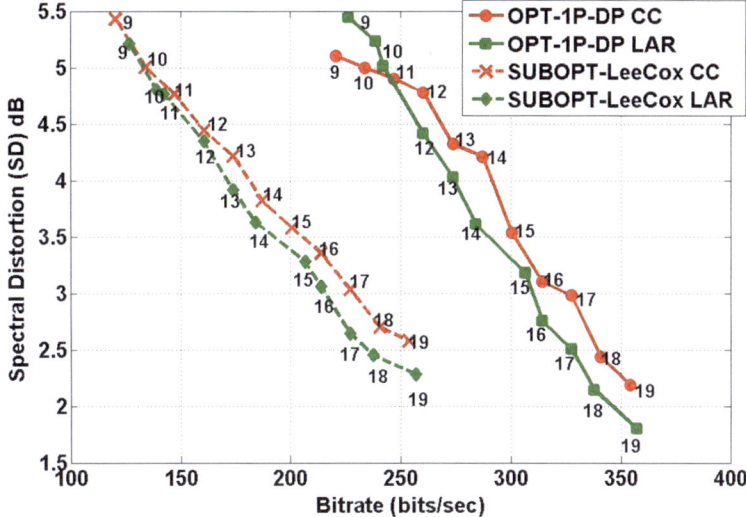

Fig. 6.9 Rate-distortion plot (SD vs. overall bit-rate) (Reused with permission from [R12])

This essentially measures how well the spectral information is quantized under the joint spectral-residual quantization scheme.

The following can be noted from Fig. 6.9:

(a) LAR performs the best since it matches only the spectra; CC performs with higher spectral distortion since it emphasizes matching both the spectra and residual and in the process compromises the spectral match alone.

(b) The spectral distortion shows a prominent reduction with bit-rate and reaches as low as 1.7–2 dB (for LAR and CC) for OPT-1P-DP for an effective overall bit-rates of 350 bits/s. (As a reference, 1 dB spectral distortion corresponds to the transparent quality [PK95]).

Since our objective is to compare the two different composite measures used under the 'no-residue transmission' framework, it is meaningful to show the 'overall' performance in terms of the average log magnitude power spectral distortion (PSD) between the input speech frames and output speech frames vs. overall bit-rate. This is shown in Fig. 6.10 for the two composite measures LAR and CC for both OPT-1P-DP (solid lines) and SUBOPT-LeeCox (dashed lines). For PSD, the single frame power spectral distortion is the squared difference between the log of the magnitude power spectra of the original speech frame and the output synthesized speech frame.

The following can be observed from Fig. 6.10:

(a) The PSD values are higher than SD values since the overall PSD includes the residual spectra mismatch (over and above LP spectral envelope differences),

(b) CC performs the best indicating that it achieves the best match of spectra and residual combined,

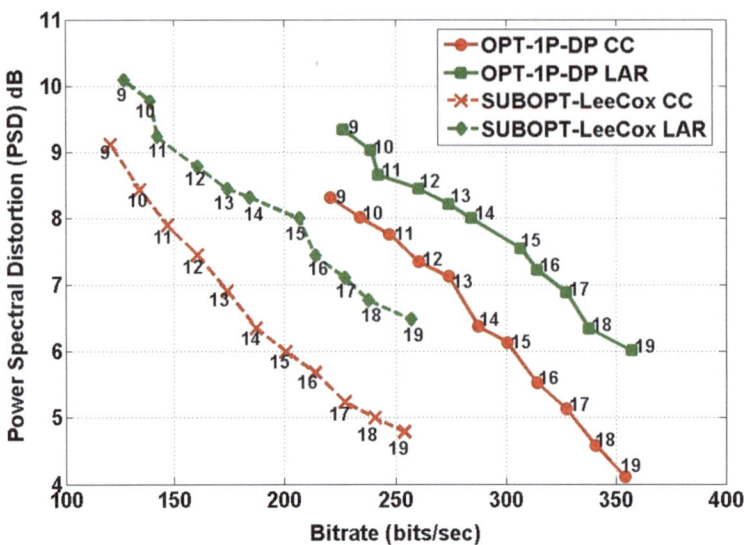

Fig. 6.10 Rate-distortion plot (PSD vs. overall bit-rate) (Reused with permission from [R12])

(c) LAR perform poorly indicating it is unable to select units with the residual also matching and points to the need for a composite measure such as CC to effectively match both spectra and residual,
(d) The other composite measures (not shown here), namely, MFCC, LAR + P and PSD fall in between these two R-D curves indicating their relative ability to match the spectra and residual jointly.

More importantly, we note from these figures that the optimality of the one-pass DP unit-selection (OPT-1P-DP) shows in the R-D performance of SD and PSD which are up to 0.5 dB lower than SUBOPT-LeeCox, particularly at the higher unit database sizes (higher bit-rates, close to 350 bits/s), clearly validating the optimality of the 1-pass DP unit-selection algorithm proposed here. Figures 6.9 and 6.10 also show OPT-1P-DP incurring an additional 100 bits/s over SUBOPT-LeeCox (as discussed in Sect. 6.2.1) with an overall bit-rate of 225–350 bits/s.

The above observations, when interpreted in a combined way (SD and PSD), corroborates with the actual quality of the synthesized speech by listening. We found that OPT-1P-DP offers an overall quality of speech which is intelligible, natural and retains speaker-identity and prosody (gain and intonation). Since OPT-1P-DP CC is 0.5 dB lower in SD and PSD than SUBOPT-LeeCox [RH09], it is closer to the input speech and offers a proportional enhancement in the overall quality of speech over SUBOPT-LeeCox: while SUBOPT-LeeCox has a perceptible coarseness representing the 2.5 dB SD, OPT-1P-DP (with its 1.7–2 dB SD), does not have this coarseness over the entire 5 s input test sentence and offers a smooth listening quality of the quantized speech. This shows how the optimality of the proposed one-pass DP based joint spectral-residual quantization manifests in the overall quality of the coded speech and the feasibility of realizing ultra low bit-rate speech coding at bit-rates of 350 bits/s within a 'no-residue-transmission' framework.

Use of better LP parameters (such as LSFs in place of LARs) and use of larger codebooks (say 20–21 bits), can take the SD values in the R-D curve of Fig. 6.9 closer to the 1 dB transparent quality reference, thereby making it feasible to realize high quality coding at ultra low bit-rates of 300–400 bits/s within the 'no-residue-transmission' framework proposed here using unit-selection based joint spectral-residual quantization scheme.

6.3 Conclusions

We have proposed an ultra low bit-rate speech coder based on unit-selection which obviates the need for transmitting any residual information by virtue of joint spectral-residual quantization under various types of composite distance measures. The system realizes spectral distortions of less than 2.5 dB at an overall rate of 250 bits/s. We have also proposed an optimal unit-selection algorithm with joint spectral-residual quantization for ultra low bit-rate speech coding which realizes a better rate-distortion performance than the former algorithm above, set in a sub-optimal unit-selection framework.

References

[A83]   B.S. Atal, Efficient coding of LPC parameters by temporal decomposition, in *Proceedings of ICASSP '83*, 1983, pp. 81–84

[BS00] E.F.M.F Badran, H. Selim, Speaker recognition using artificial neural networks based on vowel phonemes, in *Proceedings of 5th International Signal Processing, WCCC-ICSP*, vol. 2, Aug 2000, pp. 796–802

[B02]   G. Baudoin, F. Capman, J. Cernocky, F. El Chami, M. Charbit, G. Chollet, D. Petrovska-Delacrétaz, Advances in very low bit rate speech coding using recognition and synthesis techniques", in *Proceedings of the 5th International Conference on Text, Speech and Dialogue*, Brno, Czech Republic, Sept 2002, pp. 269–276

[BCC97] G. Baudoin, J. Cernocky, G. Chollet, Quantization of spectral sequences using variable length spectral segments for speech coding at very low bit rate, in *Proceedings of Eurospeech '97*, Rhodes, 1997, pp. 1295–1298

[BC03] G. Baudoin, F. El Chami, Corpus based very low bit rate speech coding, in *Proceedings of ICASSP '03*, 2003, pp. 792–795

[B99]   G. Baudoin, Speech coding at low very low bit rates, in *Proceedings of ERK '99*, vol. A, 1999, pp. 11–14

[BSH08] J. Benesty, M.M. Sondhi, Y. Huang (eds.), *Springer Handbook of Speech Processing* (Springer, Berlin, 2008)

[BA91] F. Bimbot, B.S. Atal, An evaluation of temporal decomposition, in *Proceedings of Eurospeech '91*, Sept 1991, pp. 1089–1092

[Br95] S. Bruhn, Matrix product quantization for very-low-rate speech coding, in *Proceedings of ICASSP '95*, Michigan, 1995, pp. 724–727

[B95]   F. Bimbot, R. Pieraccini, E. Levin, B. Atal, Variable-length sequence modeling: Multigrams. IEEE Signal Process. Lett. **2**(6), 111–113 (1995)

[BD84] G. Benbassat, X. Delon, Low bit rate speech coding by concatenation of sound units and prosody coding, in *Proceedings of IEEE International Conference on Acoustics, Speech, Signal Processing*, vol. 1, 1984, pp. 1.2.1–1.2.4

[CBC98a] J. Cernocky, G. Baudoin, G. Chollet, Segmental vocoder—going beyond the phonetic approach, in *Proceedings of. ICASSP '98*, vol. 2, 1998, pp. 605–608

[CBC97a] J. Cernocky, G. Baudoin, G. Chollet, Speech spectrum representation and coding using multigrams with distance, in *Proceedings of ICASSP '97*, Munich, 1997, pp. 1343–1346

[CBC97b] J. Cernocky, G. Baudoin, G. Chollet, Towards a very low bit rate segmental speech coder, *Computational models of speech pattern processing*, NATO Advanced Study Institute, 1997

[CBC00] J. Cernocky, G. Baudoin, G. Chollet, Unsupervised learning for very low bit-rate speech coding, *SCI 2000, 4th World Multiconférence on Systèmes, Cybernetics and Informatics*, vol. 6, 2000

[CBC98b] J. Cernocky, G. Baudoin, G. Chollet, Very low bit rate segmental speech coding using automatically derived units, *8th International Czech—Slovak Scientific Conference Proceedings Radioelektronika '98*, vol. 1, 1998, pp. 224–227

[CCTC97] H.C. Chen, C.Y. Chen, K.M. Tsou, O.T.-C. Chen, A 0.75 Kbps speech codec using recognition and synthesis schemes, in *Proceedings of IEEE Workshop Speech Coding Telecommunications*, 1997, pp. 27–29

[CKBC99] J. Cernocky, I. Kopecek, G. Baudoin, G. Chollet, Very low bit rate speech coding: comparison of data-driven units with syllable segments, in *Proceedings of Workshop on Text, Speech and Dialogue, TSD '99*, 1999, pp. 262–267

[C98] J. Cernocky, Speech processing using automatically derived units: Applications to very low rate coding and speaker verification, Ph.D. thesis, Universite Paris XI Orsay, Dec 1998

[CS90] M.Y. Cheng, D. O'Shaughnessy, A 450 b.p.s. vocoder with natural-sounding speech, in *Proceedings of ICASSP '90*, Apr 1990, pp. 649–652

[CS91] M.Y. Cheng, D. O'Shaughnessy, Short-term temporal decomposition and its properties for speech compression. IEEE Trans. Signal Process. **39**(6), 1282–1290 (1991)

[CS93] M.Y. Cheng, D. O'Shaughnessy, On 450-600 b/s natural sounding speech coding". IEEE Trans. Speech Audio Process. **1**(2), 207–220 (1993)

[C05] V. Chetan, Very low bit rate speech coding using segmentation, M. Tech dissertation, Department of EE, IIT-Bombay, 2005

[C08] S. Chevireddy, A syllable based segment vocoder, M.S. Thesis, Department of Computer Science, Indian Institute of Technology – Madras (IIT-M), Chennai, India, 2008

[CMC08a] S. Chevireddy, H.A. Murthy, C. Chandrasekhar, A syllable-based segment vocoder, in *Proceedings of National Conference on Communications*, Mumbai, India, 2008, pp. 442–445

[CMC08b] S. Chevireddy, H.A. Murthy, C. Chandrasekhar, Signal processing based segmentation and HMM based acoustic clustering for a syllable based segment vocoder at 1.4Kbps, *16th European Conference on Signal Processing and Communication, EUSIPCO-2008*, Lausanne, Switzerland

[CL94] P.A. Chou, T. Lookabaugh, Variable dimension vector quantization of linear predictive coefficients of speech, in *Proceedings of ICASSP '94*, 1994, pp. I-505–I-508

[C95] R.V. Cox, Speech coding standards, in *Speech Coding and Synthesis*, ed. by W.B. Kleijn, K.K. Paliwal (Elsevier, Amsterdam, 1995), pp. 49–78. Chapter 2

[DFGSL13] X. Dong, M.O. Fuyuan, C. Geng, G. Shengming, M. Li, Study of low bit rate speech codec algorithm in underwater acoustic communication. Chin J Acoustics **32**(04), 411–424 (2013)

[DP10] M. Deepak, P. Rao, Trajectory and surface modeling of LSF for low rate speech coding, in *Proceedings of NCC '10*, Chennai, India, 2010

[DW95] R.E. Donovan, P.C. Woodland, Improvements in an HMM-based speech synthesiser, in *Proceedings of Eurospeech '95*, vol. 1, 1995, pp. 573–576

[D04] S. Dusan, J. Flanagan, A. Karve, M. Balaraman, Speech coding using trajectory compression and multiple sensors, in *Proceedings of ICSLP '04*, Jeju, Korea, 2004

[D97] T. Dutoit, *An Introduction to Text-to-Speech Synthesis* (Kluwer Academic, Dordrecht, 1997)

[F93] L.J. Fransen, *Template based low data rate speech encoder* (Interim Report Naval Research Lab, Washington, DC, 1993)

References

[FC86]   E. Farges, M. Clements, Hidden Markov models applied to very low bit rate coding, in *Proceedings of ICASSP '86*, 1986, pp. 433–436

[FIS80]   J.L. Flanagan, K. Ishizaka, K.L. Shipley, Signal models for low bit-rate coding of speech. J. Acoust. Soc. Am. **68**(3), 780–791 (1980)

[F79]   J.L. Flanagan et al., Speech coding. IEEE Trans. Commun. **27**(4), 710–737 (1979)

[GTLL05]   A. Goalic, J. Trubuil, G. Lapierre, J. Labat, Real time low bit rate speech transmission through underwater acoustic channel, in *Oceans 2005*, vol. 1, June 2005, pp. 319–321

[GD96]   S. Ghaemmaghami, M. Deriche, A new approach to very low-rate speech coding using temporal decomposition, in *Proceedings of ICASSP '96*, May 1996, pp. 224–227

[GDB97a]   S. Ghaemmaghami, M. Deriche, B. Boashash, Comparative study of different parameters for temporal decomposition based speech coding, in *Proceedings of ICASSP '97*, Apr 1997, pp. 1703–1706

[GDB97b]   S. Ghaemmaghami, M. Deriche, B. Boashash, On modeling event functions in temporal decomposition based speech coding, in *Proceedings of Eurospeech '97*, Rhodes, 1997, pp. 1299–1302

[GDS98]   S. Ghaemmaghami, M. Deriche, S. Sridharan, Hierarchical temporal decomposition: a novel approach to efficient compression of spectral characteristics of speech, in *Proceedings ICSLP '98*, Sydney, 1998

[G07]   L. Girin, Long-term quantization of LSF parameters, in *Proceedings of ICASSP '07*, Honolulu, Hawaii, 2007

[H98]   W.J. Holmes, Towards a unified model for low bit-rate speech coding using a recognition-synthesis approach, in *Proceedings of 5th International Conference on Spoken Language Processing (ICSLP '98)*, Sydney, 1998

[HT04]   T. Hirai, S. Tenpaku, Using 5 ms segments in concatenative speech synthesis, in *Proceedings of 5th ISCA Speech Synthesis Workshop*, Pittsburgh, June 2004, pp. 37–42

[HR08]   D. Harish, V. Ramasubramanian, Comparison of segment quantizers: VQ, MQ, VLSQ and Unit-selection algorithms for ultra low bit-rate speech coding, in *Proceedings of ICASSP '08*, Las Vegas, Mar 2008, pp. 4773–4776

[HS92]   M. Honda, Y. Shiraki, Very low-bit-rate speech coding, in *Advances in Speech Signal Processing*, ed. by S. Furui, M.M. Sondhi (Dekker, New York, 1992), pp. 209–230

[H03]   T. Hoshiya, S. Sako, H. Zen, K. Tokuda, T. Masuko, T. Kobayashi, T. Kitamura, Improving the performance of HMM-based very low bit rate speech coding, in *Proceedings of ICASSP '03*, 2003, pp. I-800–I-803

[HB96]   A.J. Hunt, A.W. Black, Unit selection in a concatenative speech synthesis system using a large speech database, in *Proceedings of ICASSP '96*, 1996, pp. 373–376

[HN89]   Y. Hirata, S. Nakagawa, A 100bit/s speech coding using a speech recognition technique, in *Proceedings of Eurospeech '89*, 1989, pp. 290–293

[IP97]   M. Ismail, K. Ponting, Between recognition and synthesis—300 bits/second speech coding, in *Proceedings of Eurospeech '97*, 1997, pp. 441–444

[I98]   M.R. Ismail, Sub-300 bits/sec speech processing, Master's thesis, University of Surrey, Department of Electronic and Electrical Engineering, Sept 1998

[Ji12]   J. Jiang, X. Ma, Y. Li, Y. Fan, Q. Hao, A real-time underwater speech communication system, in *Proceedings of ICCASM*. Advances in Intelligent Systems Research, July 2012

[JF94]   C. Jaskie, B. Fette, A survey of low bit rate vocoders. DSP & Multimedia Technology, April 1994, pp. 26–40

[JP91]   P. Jeanrenaud, P. Peterson, 'Segment vocoder based on reconstruction with natural segments', in *Proceedings of. ICASSP '91*, vol. 1, 1991, pp. 605–608

[KF85]   G.S. Kang, L.J. Fransen, *Low-bit rate speech encoders based on Line-Spectrum Frequencies (LSF)* (Interim Report Naval Research Lab, Washington, DC, 1985)

[KCT91]   D.P. Kemp, J.S. Collura, T.E. Tremain, Multi-frame coding of LPC parameters at 600-800 bps, in *Proceedings of ICASSP '91*, 1991, pp. 609–612

[KCT92] D.P. Kemp, J.S. Collura, T.E. Tremain, LPC parameter quantization at 600, 800, 1200 bps, in *Proceedings of Tactical Communications Conference, TCC 92*, Fort Wayne, 1992, pp. 71–75

[KP95a] W.B. Kleijn, K.K. Paliwal, An introduction to speech coding, in *Speech Coding and Synthesis*, ed. by W.B. Kleijn, K.K. Paliwal (Elsevier, Amsterdam, 1995), pp. 1–48. Chapter 1

[KP95b] W.B. Kleijn, K.K. Paliwal, *Speech Coding and Synthesis* (Elsevier, Amsterdam, 1995)

[K05] A.M. Kondoz, *Digital Speech—Coding for Low Bit-Rate Communication Systems* (Wiley, Chichester, 2005)

[Luo11] J. Luo, H. Zhou, Y. Li, M. Lu, The design of underwater speech communication codec system based on AMBE-3000, in *Proceedings of International Conference on Mechatronic Science, Electric Engineering and Computer (MEC)*, Aug 2011, pp. 2471–2475

[Li11] Y. Li, J.S. Jiang, X.F. Ma, Research on ultra-low-bit-rate speech coding algorithm for underwater speech communication. Appl Mech Mater **128–129**, 247–250 (2011)

[LC01] K.S. Lee, R.V. Cox, A very low bit rate speech coder based on a recognition/synthesis paradigm. IEEE Trans. Speech Audio Process. **9**(5), 482–491 (2001)

[LC02] K.S. Lee, R.V. Cox, A segmental speech coder based on a concatenative TTS. Speech Commun. **38**, 89–100 (2002)

[LC99] K.-S. Lee, R.V. Cox, TTS based very low bit rate speech coder, in *Proceedings of IEEE International Conference on Acoustics, Speech, Signal Processing*, 1999, pp. 181–184

[Lee05] M.E. Lee, A.S. Durey, E. Moore, M. Clements, Ultra low bit rate speech coding using an ergodic hidden Markov model, in *Proceedings of ICASSP '05*, 2005, pp. I-765–I-768

[LF93] J.M. Lopez-Soler, N. Farvardin, A combined quantization-interpolation scheme for very low bit rate coding of speech LSP parameters, in *Proceedings of ICASSP '93*, vol. 2, 1993, pp. 21–24

[M06] J. Makhoul, Speech processing at BBN. IEEE Ann. Hist. Comput. Arch. **28**(1), 32–45 (2006)

[MCRK77] J. Makhoul, C. Cook, R. Schwartz, D. Klatt, A feasibility study of very low rate speech compression system, Report no. 3508, Bolt Beranek and Newman Inc., Cambridge, MA, Feb 1977

[M85] J. Makhoul et al., Vector quantization in speech coding. Proc. IEEE **73**(11), 1551–1588 (1985)

[MD91] J.P. Martens, L. Depuydt, Braod phonetic classification and segmentation of continuous speech by means of neural networks and dynamic programming. Speech Commun. **10**(1), 81–90 (1991)

[MTK98] T. Masuko, K. Tokuda, T. Kobayashi, A very low bit rate speech coder using HMM with speaker adaptation, ICSLP 1998

[MTKI97] T Masuko, K Tokuda, T Kobayashi, S Imai, Voice characteristics conversion for HMM-based speech synthesis, in *Proceedings of ICASSP '97*, vol. III, 1997, pp. 1611–1614

[M08] A.V. McCree, Low-bit-rate speech coding, in *Springer Handbook of Speech Processing*, ed. by J. Benesty, M.M. Sondhi, Y. Huang (Springer, Berlin, 2008), pp. 331–350. Chapter 26

[MB95] A.V. McCree, T.P. Barnwell III, A mixed excitation LPC vocoder model for Low Bit rate speech coding. IEEE Trans. Speech Audio Process. **3**(4), 242–250 (1995)

[MBCC01] P. Motlicek, G. Baudoin, J. Cernocky, G. Chollet, Minimization of transition noise and HNM synthesis in very low bit rate speech coding, in *Text, Speech and Dialogue, Proceedings of 4th International Conference, TSD 2001*, Zelezna Ruda, Czech Republic, Springer, Sept 2001, pp. 305–312

[MGC01] P. Motlıcek, G. Baudoin, J. Cernocky, Diphone-like units without phonemes—option for very low bit rate speech coding, in *Proceedings of IEEE—EUROCON '2001*, Slovakia, Jul 2001, pp. 463–466

[MY11] H.A. Murthy, B. Yegnanarayana, Group delay functions and its applications in speech technology. Sadhana **36**(Part 5), 745–782 (2011)

[ML92] B. Mouy, P. de La Noue, Voice transmission at a very low bit rate on a noisy channel: 800 bps vocoder with error protection to 1200 bps, in *Proceedings of ICASSP '92*, vol. II, San Francisco, 1992, pp. 149–152

[MLG95] B. Mouy, P. de La Noue, G. Goudezeune, NATO STANAG 4479: a standard for an 800 bps vocoder and channel coding in HF-ECCM system, in *Proceedings of ICASSP '95*, 1995, pp.480–483

[McC06] A. McCree, A scalable phonetic vocoder framework using joint predictive vector quantization of MELP parameters, in *Proceedings of ICASSP '06*, vol. I, 2006, pp. 705–708

[NA05] S. Narayanan, A. Alwan, *Text to speech synthesis: new paradigms and advances* (Pearson Education, Upper Saddle River, 2005)

[NM04] T. Nagarajan, H.A. Murthy, Group delay based segmentation of spontaneous speech into syllable-like units. EURASIP J. Appl. Signal Process. **17**, 2614–2625 (2004)

[N84] H. Ney, The use of one-stage dynamic programming algorithm for connected word recognition. IEEE Trans. Acoust. Speech Signal Process. **32**(2), 263–271 (1984)

[NE04] F. Nordén, T. Eriksson, Time evolution in LPC spectrum coding. IEEE Trans. Speech Audio Process. **12**(3), 290–301 (2004)

[OPT00] M.J. Ovens, K.M. Ponting, M.E. Turner, Ultra low bit rate voice coding, Technical report, 20/20 Speech Ltd, UK, 2000

[PCB04b] M. Padellini, F. Capman, G. Baudoin, Dynamic unit selection for very low bit rate coding at 500 bits/sec, *Text, Speech and Dialogue (TSD '04), Lecture Notes in Computer Science*, Springer, Berlin, 2004, pp. 417–423

[PCB04a] M. Padellini, F. Capman, G. Baudoin, Very low bit rate (VLBR) speech coding around 500 bits/sec, *EUSIPCO 2004: (XII. European Signal Processing Conference)*, Vienna, Austria, Sept 2004, pp. 1669–1672

[PCB05] M. Padellini, F. Capman, G. Baudoin, Very low bit rate speech coding in noisy environments, *SPECOM '2005—10th International Conference on Speech and Computer*, 2005

[PA93] K.K. Paliwal, B.S. Atal, Efficient vector quantization of LPC parameters at 24 bits/frame. IEEE Trans. Speech Audio Process. **1**, 3–14 (1993)

[PK95] K.K. Paliwal, W.B. Kleijn, Quantization of LPC parameters, in *Speech Coding and Synthesis*, ed. by W.B. Kleijn, K.K. Paliwal (Elsevier, Amsterdam, 1995), pp. 433–466. Chapter 12

[PC92] S. Parthasarathy, C.H. Coker, On automatic estimation of articulatory parameters in text-to-speech system. Comput. Speech Lang. **6**(1), 37–75 (1992)

[Pepp91] D. Pepper, M. Clements, On the phonetic structure of a large hidden Markov model, in *Proceedings of ICASSP '91*, 1991, pp. 465–468

[Pepp90] D.J. Pepper, Large hidden Markov model state interpretation as applied to automatic phonetic segmentation and labeling, Ph.D. dissertation, Georgia Institute of Technology, 1990

[PJV90] P. Peterson, P. Jeanrenaud, J. Vandegrift, Improving Intelligibility of a 300 b/s Segment Vocoder, in *Proceedings of ICASSP '90*, 1990, pp. 653–656

[P95] L. Piazzo, A new very low bit rate speech coder: The step decomposition vocoder, in *Proceedings of Eurospeech '95*, Madrid, 1995, pp. 237–240

[PD89] J. Picone, G.R. Doddington, A phonetic vocoder, in *Proceedings of ICASSP '89*, 1989, pp. 580–583

[PSVH10] A. Pradhan, S. Chevireddy, K. Veezhinathan, H. Murthy, A low-bit rate segment vocoder using minimum residual energy criteria, in *Proceedings of National Conference on Communication (NCC-2010)*, Chennai, India, 2010, pp. 246–250

[PGV97] P. Prandoni, M. Goodwin, M. Vetterli, Optimal time segmentation for signal modeling and compression, in *Proceedings of ICASSP*, vol. 3, Munich, Apr 1997, pp. 2029–2032
[PV00] P. Prandoni, M. Vetterli, R/D optimal linear prediction. IEEE Trans. Speech Audio Process. **8**(6), 646–655 (2000)
[RJ93] L.R. Rabiner, B.H. Juang, *Fundamentals of Speech Recognition* (Prentice Hall, Englewood Cliffs, NJ, 1993)
[RWJ86] L.R. Rabiner, J.G. Wilpon, B.-H. Juang, A segmental K-means training procedure for connected word recognition. AT&T Bell Lab. Tech. J. **65**(3), 21–31 (1986)
[RH07] V. Ramasubramanian, D. Harish, An optimal unit-selection algorithm for ultra low bit-rate speech coding, in *Proceedings of ICASSP '07*, Hawaii, Apr 2007, pp. IC-541–IC-544
[RH06] V. Ramasubramanian, D. Harish, An unified unit-selection framework for ultra low bit-rate speech coding, in *Proceedings of Interspeech '06*, Pittsburgh, Sept 2006, pp. 217–220
[RH08] V. Ramasubramanian, D. Harish, Low complexity near-optimal unit-selection algorithm for ultra low bit-rate speech coding based on N-best lattice and Viterbi decoding, in *Proceedings of Interspeech '08*, Brisbane, Sept 2008, p. 44
[RH09] V. Ramasubramanian, D. Harish, Ultra low bit-rate speech coding based on unit-selection with joint spectral-residual quantization: No transmission of any residual information, in *Proceedings of Interspeech '09*, Brighton, Sept 2009, pp. 2615–2618
[RSS03] V. Ramasubramanian, A.K.V. Sai Jayram, T.V. Sreenivas, Language identification using parallel sub-word recognition—an ergodic HMM equivalence, in *Proceedings of Eurospeech '03*, Geneva, Switzerland, Sept 2003, pp. 1357–1360
[RS04] V. Ramasubramanian, T.V. Sreenivas, Automatically derived units for segment vocoders, in *Proceedings of ICASSP '04*, Montreal, Canada, 2004, pp. 473–476
[R12] V. Ramasubramanian, Ultra low bit-rate speech coding: An overview and recent results, *IEEE SPCOM*, Indian Institute of Science, Bangalore, 2012
[RT96] C.M. Ribeiro, I.M. Trancoso, Application of speaker modification techniques to phonetic vocoding, in *Proceedings of ICSLP '96*, 1996, pp. 306–309
[RTC00] C.M. Ribeiro, I.M. Trancoso, D.A. Caseiro, Phonetic vocoder assessment, in *Proceedings of ICSLP '00*, 2000, pp. 830–833
[RT98] C.M. Ribeiro, I.M. Trancoso, Improving speaker recognisability in phonetic vocoders, in *Proceedings of 5th ICSLP '98*, paper 0448, Sydney, 1998
[RT97] C.M. Ribeiro, I.M. Trancoso, Phonetic vocoding with speaker adaptation, in *Proceedings of Eurospeech '97*, vol. 3, 1997, pp. 1291–1294
[RSM82b] S. Roucos, R. Schwartz, J. Makhoul, Segment quantization for very-low-rate speech coding, in *Proceedings of ICASSP*, vol. 3, Paris, France, 1982, pp. 1565–1568
[RSM82a] S. Roucos, R.M. Schwartz, J. Makhoul, Vector quantization for very-low-rate coding of speech, in *Proceedings of IEEE Globecom '82*, Miami, FL, 1982, pp. 1074–1078
[RWR87] S. Roucos, A. Wilgus, W. Russel, A segment vocoder algorithm for real-time implementation, in *Proceedings of ICASSP '87*, 1987, pp. 1949–1952
[RW85] S. Roucos, A.M. Wilgus, The waveform segment vocoder: a new approach for very-low-rate speech coding, in *ICASSP '85*, 1985, pp. 236–239
[RSM83] S. Roucous, R.M. Schwartz, J. Makhoul, A segment vocoder at 150 b/s, *Proceedings of ICASSP '83*, Boston, 1983, pp. 61–64
[SRS02] A.K.V. Sai Jayram, V. Ramasubramanian, T.V. Sreenivas. Robust parameters for automatic segmentation of speech. in *Proceedings of ICASSP*, May 2002, pp. I-513–I-516
[S79] H. Sakoe, Two level DP-matching—a dynamic programming based pattern matching algorithm for connected word recognition. IEEE Trans. Acoust. Speech Signal Process. **ASSP-27**(6), 588–595 (1979)

[SR05]	S.A. Santosh Kumar, V. Ramasubramanian, Automatic language identification using ergodic-HMM, in *Proceedings of ICASSP '05*, Philadelphia, Mar 2005, pp. I-609–I-612
[SR83]	R. M. Schwartz, S. E. Roucos, A comparison of methods for 300-400 b/s vocoders, in *Proceedings of ICASSP'83*, Boston, 1983, pp. 69–72
[SS92]	J. Schroeter, M.M. Sondhi, Speech coding based on physiological models of speech production, in *Advances in Speech Signal Processing*, ed. by S. Furui, M.M. Sondhi (Dekker, New York, 1992), pp. 231–268
[SKMKZ79]	R. Schwartz, J. Klovstad, J. Makhoul, D. Klatt, V. Zue, Diphone synthesis for phonetic vocoding, in *Proceedings of ICASSP*, April 1979, pp. 891–894
[SKMS80]	R. Schwartz, J. Klovstad, J. Makhoul, J. Sorensen, A preliminary design of a phonetic vocoder based on a diphone model, in *Proceedings of ICASSP '80*, 1980, pp. 32–35
[SH88]	Y. Shiraki, M. Honda, LPC speech coding based on variable-length segment quantization". IEEE Trans. Acoust. Speech Signal Process. **36**(9), 1437–1444 (1988)
[SS05]	M.M. Sondhi, D.J. Sinder, Articulatory modeling: a role in concatenative text to speech synthesis, in *Text to Speech Synthesis: New Paradigms and Advances*, ed. by S. Narayanan, A. Alwan (Pearson Education, Prentice Hall, 2005), pp. 85–109. Chapter 4
[Sp94]	A.S. Spanias, Speech coding: a tutorial review. Proc. IEEE **82**(10), 1541–1582 (1994)
[Sung98]	Sung Joo Kim, Sangho Lee, Woo Jin Han, Yung Hwan Oh, Efficient quantization of LSF parameters based on temporal decomposition, in *Proceedings of ICSLP '98*, Sydney 1998
[S94]	T. Svendsen, Segmental quantization of speech spectral information, in *Proceedings of ICASSP '94*, 1994, pp. I-517–I-520
[SS87]	T. Svendsen, F.K. Soong, On the automatic segmentation of speech signals, in *Proceedings of ICASSP '87*, 1987, pp. 77–80
[T98]	K. Tokuda, T. Masuko, J. Hiroi, T. Kobayashi, T. Kitamura, A very low bit rate speech coder using HMM-based speech recognition/synthesis techniques, in *Proceedings of ICASSP '98*, 1998, pp. 609–612
[T13]	K. Tokuda, Y. Nankaku, T. Toda, H. Zen, J. Yamagishi, K. Oura, Speech synthesis based on hidden Markov models. Proc. IEEE **101**(5), 1234–1251 (2013)
[T82]	T.E. Tremain, The government standard linear predictive coding algorithm: LPC-10, *Speech Technology*, Apr 1982, pp. 40–49
[TG85]	C. Tsao, R.M. Gray, Matrix quantizer design for LPC speech using the generalized Lloyd algorithm. IEEE Trans. ASSP **33**(3), 537–545 (1985)
[TKCK93]	T. Tremain, D. Kemp, J. Collura, M. Kohler, Evaluation of low rate speech coders for HF, in *Proceedings of ICASSP '93*, vol. II, 1993, pp. 163–166
[VB97]	P. Vepyek, A.B. Bradley, Consideration of processing strategies for very-low-rate compression of wide band speech signal with known text transcription, in *Proceedings of Eurospeech '97*, vol. 3, 1997, pp. 1279–1282
[VM89]	A.M. Van Dijk-Kappers, S.M. Marcus, Temporal decomposition of speech. Speech Commun. **8**(2), 125–135 (1989)
[V89]	A.M. Van Dijk-Kappers, Temporal decomposition of speech and its relation to phonetic information, Ph.D. thesis, Technische Universiteit Eindhoven, Eindhoven, 1989
[v91]	J.P. van Hemert, Automatic segmentation of speech. IEEE Trans. Acoust. Speech Signal Process. **39**(4), 1008–1012 (1991)
[vHvB95]	V.J. van Heuven, R. van Bezooijen, Quality evaluation of synthesized speech, in *Speech Coding and Synthesis*, ed. by W.B. Kleijn, K.K. Paliwal (Elsevier, Amsterdam, 1995), pp. 707–738. Chapter 21

[VM90] E. Vidal, A. Marzal, A review and new approaches for automatic segmentation of speech signals, in *Signal Processing V: Theories and Applications*, ed. by L. Torres, E. Masgrau, M.A. Lagunas (Elsevier Science Publisher B.V, Amsterdam, 1990), pp. 43–53
[WJC83] D.Y. Wong, B.H. Juang, D.Y. Cheng, Very low data rate speech compression with LPC vector and matrix quantization, in *Proceedings of ICASSP '83*, Boston, 1983, pp. 65–68
[W82] D.Y. Wong et al., An 800 b/s vector quantization LPC vocoder. IEEE Trans. ASSP **30**(6), 770–780 (1982)
[W91] C.J. Weinstein, Opportunities for advanced speech processing in military computer-based systems. Proc. IEEE **79**(11), 1626–1641 (1991)

MIX
Papier aus verantwortungsvollen Quellen
Paper from responsible sources
FSC® C105338

If you have any concerns about our products,
you can contact us on
ProductSafety@springernature.com

In case Publisher is established outside the EU,
the EU authorized representative is:
**Springer Nature Customer Service Center GmbH
Europaplatz 3, 69115 Heidelberg, Germany**

Printed by Libri Plureos GmbH
in Hamburg, Germany